U0345807

能量与热力学建筑书系
Energy & Thermodynamic Architecture

# 热力学建筑原型

# Thermodynamic Architectural Prototype

李麟学 著
LI Linxue

同济大学出版社
TONGJI UNIVERSITY PRESS
中国·上海

图书在版编目（CIP）数据

热力学建筑原型 / 李麟学著 .— 上海：同济大学出版社，2019.3
（能量与热力学建筑书系）
ISBN 978-7-5608-6148-7

Ⅰ.①热… Ⅱ.①李… Ⅲ.①热力学 – 应用 – 建筑学 – 研究 Ⅳ.① TU-05

中国版本图书馆 CIP 数据核字（2015）第 320155 号

同济大学人居环境生态与节能联合研究中心资助（项目名称：能量形式化与热力学建筑前沿理论建构）
国家自然科学基金资助（项目批准号 51278340）
上海市科学技术委员会 2018 年度“科技创新行动计划”国际学术合作交流项目资助（项目编号 18230722500，项目名称：环境性能与热力学导向的崇明生态岛低能耗建筑整合设计研究）

# 热力学建筑原型

李麟学　著

出 品 人：华春荣
策　　划：秦蕾 / 群岛工作室

责任编辑：晁　艳
麟和团队：陶思旻、何美婷、侯苗苗、张琪
责任校对：徐春莲
平面设计：张　微
版　　次：2019 年 3 月第 1 版
印　　次：2019 年 3 月第 1 次印刷
印　　刷：上海安枫印务有限公司
开　　本：720mm×1000mm　1/16
印　　张：16
字　　数：320 000
书　　号：ISBN 978-7-5608-6148-7
定　　价：98.00 元
出版发行：同济大学出版社
地　　址：上海市四平路 1239 号
邮政编码：200092
网　　址：http://www.tongjipress.com.cn
经　　销：全国各地新华书店



*Thermodynamic Architectural Prototype*
by: LI Linxue
ISBN 978-7-5608-6148-7

Publisher: HUA Chunrong
Initiator: QIN Lei / Studio Archipelago
Editor: CHAO Yan
ATELIER L+ Team: TAO Simin, HE Meiting, HOU Miaomiao, ZHANG Qi
Proofreader: XU Chunlian
Graphic Designer: ZHANG Wei

Published in March 2019, by Tongji University Press,
1239, Siping Road, Shanghai 200092, China.
www.tongjipress.com.cn

# 形式追随热力学

## 李振宇

“形式追随功能”这句话大家耳熟能详。如果回望历史，我们可以发现，原来在不同的历史时期，形式会追随不同的东西；同一时期，也会有人让形式追随不同的东西。比如我们可以说，上古时期，形式主要追随生存；后来，形式主要追随秩序；到了现代主义建筑，形式追随功能成为主流；现代主义之后，形式追随各种各样的多元；再后来，形式追随生态也成为许多建筑师追求的目标，例如形式追随气候，形式追随本土，等等。今天李麟学教授出版的《热力学建筑原型》一书，就是“形式追随热力学”的呈现。

从“追随生存”到“追随生态”，建筑学的发展既包括原有传统设计领域的本体演变，又包括了新领域的不断拓展。这种拓展既是建筑类型学的贡献，也是不断引入新知识体系的结果。李麟学教授2014年在哈佛大学做访问学者，参与开展热力学建筑方向的国际联合研究与教学，收获颇丰。之后，结合哈佛大学与同济大学的合作研究，以同济大学建筑学四年级的设计自选课为载体，他带领学生展开了以“热力学建筑原型”为主题的教学实验。我们欣喜地看到，这一领域的研究既是对生态建筑、绿色建筑知识的拓展，又结合了建筑本体设计的前沿视野，提供了应对建成环境的新方法、新策略与新工具。这一专业设计领域的探索展现了建筑设计研究与教学的新边界。正如李麟学教授在书中提出的，如何将对建筑本体的设计关注与环境挑战结合起来，成为本书展现的学生课程设计的主要着力点。

共同推动这一领域研究与教学的哈佛大学的伊纳吉·阿巴罗斯教授，在热力学建筑方向为建筑设计打开了全新的视野，同时他也是同济大学建筑与城市规划学院“国家高等学校学科创新引智计划”（简称“111计划”）中“未来城市与未来建筑”项目的领衔教授。参与本书的其他老师，也是国内外此领域的前沿实践者，本书成果充分体现了同济大学国际化教学的追求和成果。

“热力学建筑原型”体现了跨学科教学的探索，通过整合建筑设计、建筑技术、环境性能、建筑材料，以及来自能源和环境专业的知识，提供了建筑学教学的新尝试；学生的设计成果展现出跨学科研究与教学的巨大潜力。教学的组织遵循“原型生成—环境性能—场地植入—热力学物质化”的清晰思路，展现了教学团队更高的追求。跨学科教学也是同济大学建筑与城市规划学院在教学探索方面的新方向，本书展示的成果，为我们提供了一个令人振奋的开端。

我们期待，随着“热力学建筑”这一建筑学研究与教学新领域的开启，会出现更多的创新教学实验、理论研讨与设计实践！

# Form Follows Thermodynamics

## LI Zhenyu

"Form follows Function" is a well-known phrase. Looking back, we may discover that form follows different objects in different historical periods, and sometimes, it follows different objects even in the same historical period. For example, it can be said that form followed survival in ancient times. Afterwards, form mainly followed order. When it came to the period of modern architecture, it became a mainstream that form followed function. Then, form followed various objects - ecology, climate, and native culture. This book, *Thermodynamic Architectural Prototype,* written by Professor Li Linxue is typical of "Form follows Thermodynamics".

From "following survival" to "following ecology", the development of architecture evolves not only in the traditional design fields but also in the expansion of new fields. Such expansion includes contributions to architectural typology and continuous introduction of new knowledge systems. Professor Li participated in an international collaborative research and teaching program towards thermodynamic architecture when he was visiting Harvard in 2014. Later, in line with the collaboration between Tongji and Harvard, he led the teaching experiment themed "Thermodynamic Architectural Prototype" as an optional course for senior students majored in architecture in Tongji University. We are excited to see that this new field has not only extended our knowledge of ecological architecture and green architecture but also combined a cutting-edge view of architecture design, offering new methods, strategies, and tools to tackle the problems of the built environment. By exploring this new design field, a new frontier of research and the teaching of architectural design is unfolded. As Professor Li mentioned, the focus of this book is how to combine the architectural design with environmental challenges.

Professor Iñaki Ábalos from Harvard who promoted the research and teaching of this field with us is the chief professor of the project of "Future City and Future Architecture", a "111 national talent introduction base" initiated by College of Architecture & Urban Planning of Tongji University. His innovations in thermodynamic architecture provide new perspectives for architectural design. Other contributors of this book are also leading practitioners in this field at home and abroad. The book reflects how far we have gone on the way of international teaching.

*Thermodynamic Architectural Prototype* is an exploration of interdisciplinary teaching, a new attempt to architectural teaching by combining architectural design, architectural technologies, environmental performance, building materials, and expertise in energy and environment. The design works by students prove the great potential of interdisciplinary research and teaching. The organization of teaching follows the steps of "prototype generation", "environmental performance", "field implantation" and "thermodynamic materialization". This book is an encouraging start for us to continue exploring interdisciplinary teaching at the College of Architecture & Urban Planning of Tongji University.

With the groundbreaking efforts made by Professor Li and his team in "Thermodynamic Architecture", we have every reason to expect more creative teaching experiments, theoretical discussion, and design activities.

# “迭代理论”视角下的热力学理论模型

李翔宁

20世纪以来的建筑和城市理论从来没有在某种建筑理论和城市模型面前徘徊不前。相反，从后现代建筑理论、文脉主义、历史主义、类型学、现象学、人类学理论，到计算机辅助设计、绿色(生态)建筑、海绵城市、智慧城市，再到人工智能和大数据城市，历史和理论教科书不断增添新的篇章，新的建筑和城市分析研究的模型也不断涌现。来自计算机科学的迭代理论告诉我们，迭代是一种重复反馈的过程，每一次过程的重复称为一次迭代，而每一次迭代得到的结果会作为下一次迭代的初始值。

基于这种理论，今天我们可以看到科技和互联网产品1.0、2.0、3.0不同版本的不断攀升，而“迭代”的“代”字似乎也被赋予了一代一代更替的意义。今天，我们不再追求某种理论的终极完美，因为这种终极完美的状态并不一定存在，或者说，我们追求的正是理论或产品的不断演进更替，而一旦成为最终一代产品，不再向前发展，也许就是这个产品寿终正寝之时。

长期以来，困扰建筑和城市设计领域的设计者与研究者的问题之一，就是如何达成建筑形态和建筑的物理、技术性能的一致性：形式设计的大师遭受形式主义的诟病，而各种生态节能建筑则被抨击为形态粗拙生硬。基于这样的现实，哈佛大学的伊纳吉·阿巴罗斯和基尔·莫教授共同开展了跨越建筑设计、历史与技术的“热力学”研究，在20世纪经典现代主义以来的设计中针对热力学建筑性能进行考察，进而发展出当代设计的一整套方法论，通过对建筑中能量流动机理的科学分析，为建筑本体诸如形式、功能、空间等的组织提供支撑。

两年前，伊纳吉·阿巴罗斯教授关于热力学研究方法论的著作中文版在同济大学出版社出版，他邀请我为该书作序，我意识到这是认识当代城市和建筑的一次理论跃迁，热力学建筑的模型是既有建筑理论模型的一次迭代，它实际上在建筑本体设计和能量之间建立起一座桥梁。随后阿巴罗斯教授在哈佛的研究生设计教学之中贯彻了热力学建筑方法的模式，并不断检验和修正热力学建筑方法。

同济大学李麟学教授将阿巴罗斯教授的热力学建筑方法引入两校的联合设计教学，一方面引入当代中国乃至亚洲地域内城市与建筑的具体案例进行热力学模型的迭代分析，另一方面在阿巴罗斯教授找到的工具的基础上，试图更精确地在设计与技术之间进行定位，共同探讨该方法的场域、界限与潜能，在热力学建筑理论内部不断完成修正、更新与迭代。我想，阿巴罗斯教授在同济参与的教学活动对他完善和升级自己的热力学建筑方法一定也有启发。当然绝对的完美是不存在的，这个过程对于一种理论模型自身的发展和进步具有至关重要的意义。这些实践使得热力学建筑能够超越简单的基础设施、结构、能量的物理性能计算而关注物质的本真意义，以一种更加深刻的视角进行环境的调控，并借以观照我们自身的身体体验和建造文化。或者用李麟学教授的话说——“以技术文化的观念重构建筑思考的整体性”。这或许正是大家面前的这本书能够为我们提供的启示。

# The Thermodynamic Theoretical Model from the Perspective of "Iteration Theory"

LI Xiangning

Since the beginning of the twentieth century, the development of architectural and urban theories has never stopped by a certain architectural theory and urban model. On the contrary, the textbooks of history and theories have been made thicker and thicker by chapters ranging from post-modern architectural theory, contextualism, historicism, typology, phenomenology, anthropological theory, to computer aided design, green (ecological) architecture, sponge cities, smart cities, to artificial intelligence and big data cities. Moreover, new architectures and new models for urban analysis and research are also emerging. It can be told from the iteration theory originating from computer science that Iteration is a process of repetitive feedback. Each repetition of a process is called an Iteration, and the result of each iteration is then taken as the initial value for the next iteration.

Today, an application of this theory lets us see the version numbers of technology and Internet products constantly rising from 1.0, 2.0 to 3.0. Meanwhile, the word "iteration" seems to carry the meaning of replacement, one generation after another. Today, we are no longer pursuing the ultimate perfection of a certain theory, because this ultimate perfect state does not necessarily exist; alternatively, what we are after becomes the constant evolution of theory or product. Once a model turns out to be the final generation of a product without any plan for further development, it may come the time when this product reaches the end of its life cycle.

For a long time, one problem troubling the designers and researchers in the field of architecture and urban design is how to achieve the consistency between the architectural form of a building and its physical and technical performance: masters of formal design are condemned for formalism while various eco-friendly and energy-efficient buildings are criticized for being rough and rigid. To address this situation, Prof. Iñaki Ábalos and Prof. Kiel Moe (both of whem are from Harvard University) jointly conducted a "thermodynamic" study whose scope covers the design, history, and technology aspects of architecture. A complete set of contemporary design methodologies has been developed from the examinations of thermodynamic architecture in the Classic Modernism Design since the twentieth century. The scientific analysis of the energy flow mechanism inside a building also supports to organize the ontological form, function, and space of that building.

Two years ago, the Chinese version of Prof. Iñaki Ábalos' thermodynamics research methodology was published by the Tongji University Press. Invited by Prof. Ábalos to preface that book, I realized that it is a theoretical leap for us to understand the contemporary architecture and urbanism. The model of thermodynamic architecture is not only an iteration of the existing architectural theories and models but also a bridge that actually connects the ontological design and the energy flow of a building. Later on, Prof. Ábalos started to implement such methodology of thermodynamic architecture into the graduate-level designing courses at Harvard and continues to test and refine this methodology.

Professor LI Linxue at Tongji University has introduced Prof. Iñaki Ábalos' methodology of thermodynamic architecture to the designing program jointly offered by both universities. On one hand, some actual cases of cities and buildings in China and even in the entire Asian region have been subjected to the iterative analysis by this thermodynamic model; on the other hand, based on the tools developed by Prof. Ábalos, efforts have been made to more accurately find a position between design and technology and to collaboratively explore the scope, boundary, and potential of such method. Refinement, updating, and iteration are continuously performed within the theory of thermodynamic architecture. I believe that Prof. Ábalos' teaching activities at Tongji can also inspire him to further develop and improve his methodology of thermodynamic architecture. It is no doubt that there is no such thing as absolute perfection; the significance of this process, however, is its critical importance to the development and improvement of a theoretical model itself. Those practices enable thermodynamic architecture to go beyond the simple calculations of certain physical properties like infrastructure, construction, and energy to focus on the true meaning of the matter, to regulate the environment in a more profound perspective, and to observe our own physiological experience and building culture. In other words, as pointed out by Prof. LI Linxue, to "reconstruct the integrity of architectural thinking through the concept of technical culture". This may be the revelation that this book in front of us can offer.

# 能量与热力学建筑书系

《热力学建筑视野下的空气提案——设计应对雾霾》

《热力学建筑原型》

《能量与热力学建筑前沿》

《环境智能建筑前沿》

《建筑自然系统论》

《自然系统建构——麟和建筑设计实践》

# 目录

## 序



## 主题文章



## 设计成果

## Energy & Thermodynamic Architecture

# Contents

## PREFACES



## ESSAYS



## WORKS

# 主题
# 文章

# ESSAYS

# 热力学考古：建筑设计方法论革新

李麟学

能量与热力学建筑在当代能够被重视，相关研究被重新发掘，其背景之一是全球化与广泛的可持续语境下建筑学的内在危机。正如美国建筑理论家萨拉·怀汀所讲："当代建筑日渐关注诸如生态和社会责任等议题，传统建筑学和理论的自主性正受到挑战。"当我们重新关注建筑中的能量议题与热力学建筑，必然涉及对一个知识体系的考古与重构。将热力学定律应用于建筑领域，"空气"便成为空间组织的主角；建筑可理解为一种物质的组织，并由这种组织带来 "能量流动"的秩序，同时平衡与维持建筑的"物质形式"。

从 2015 年开始，同济大学建筑与城市规划学院开展了一系列热力学建筑的研讨与教学活动，而四年级一个完整学期的自选设计课程"热力学建筑原型"则成为理论研讨与设计教学紧密结合的一个平台。在最初的课程框架设计中，知识生产和设计探索是教学的两个支点。对能量与热力学的知识系统考古，对自然、系统、控制论、生态学、气象学、热力学等跨学科知识的借鉴，是设计课程的必备环节，在一个更广阔背景下的知识生产成为形式设计与操作的深层逻辑与支撑。正如基尔·莫所讲："最终，与其把建筑的热力学粗暴理解为抽象的能量分析和优化，不如将它作为另一种选择的基本准则——一种使建成环境政治更具有世界主义色彩的选择。"

### 环境调控与作为开放系统的建筑

雷纳·班纳姆在《可控环境的建筑》一书中展示了机械工业时代建筑师面临的环境调控课题与设计策略，在指出现代建筑越来越依靠技术设备的同时，他发现"在正确的情况下，一种真正的精密处理人与环境的系统方法未必要依靠复杂的机械化"。正如他对于赖特设计的贝克之家的评论，赖特对机器设备的考虑不仅是合理地安装它们，更重要的是让它们与建筑结构、功能等合作，从而让整体的功效优于各部分之和。

被伊纳吉·阿巴罗斯称为"21 世纪的班纳姆"的基尔·莫，对于现代建筑以来的环境调控持强烈的批判意识。他在《隔离的现代主义》中反思了自现代主义以来的建筑调控与隔离传统，指出其技术体系与"热力学法则"的背离，呼吁对建筑"隔离系统"进行重新评价。这种隔离以空调使用带来的"人工气候"和不断强化的建筑密闭隔热为特征，基尔·莫称之为"作为冰箱的建筑"。在 20 世纪 40 年代空调普及之前，遍布世界各地的是传统性的前现代"气候性建筑"，后被现代主义的"气密性建筑"所替代，而今天以"能量评级认证"为代表的节能建筑，某种程度上依然是这一"环境隔离"传统的延续和强化。这种环境隔离成为大量能源消耗和环境危机的诱因之一，同时在建筑室内层面，空调带来的空间同质化造就了库哈斯所称的"垃圾空间"的蔓延，建筑本身丰富的建构逻辑也常常成为设备装修的牺牲品；而在更大的城市与生态层面，封闭建筑系统对于"内部效率"的强化，则以全生命周期更大的能量消耗以及城市空间活力的丧失作为代价。

根据伊利亚·普里高津的耗散结构理论，建筑是一个"开放的非平衡系统"，是一个热力学"耗散结构”，这一耗散体以最大化的能量交换和熵的维持为特征，且必须在一个整体的热力学系统中加以考量。将热力学科学的知识体系引入建筑领域，带来了新的但又具有考古意味的建筑环境调控范式。在艾德里安·比朗的系统图解中，开放系统带来的是一种结构性的根本改变。热力学建筑不是建筑的“附加式”呈现，也并非仅仅把能源使用效率和节能作为目标，而是基于一种“结构化”的目标，正如德勒兹所称的“动力造就形式”，将物质能量作为设计的动力之源，探究建筑作为一个开放系统所具有的本体设计潜力，通过在能量、物质、形式之间深入探究，使得热力学在触及建筑“自主性”的同时还具备掌握环境性能的“工具性”，从而达到一种“自主的工具性”。从这个意义上讲，热力学建筑试图建构的是一种形式法则，为建筑环境调控提供新的范式，从而在能量运作原理的层面将可持续议题带入建筑视野，这不仅仅是一个技术的，同时也是批判的、文化的、政治的建筑议程。

## 能量流动机理与图解

将能量这一议题重新作为建筑学研究的重点，用热力学的方法思考建筑，也就是将建筑作为能量的热力学容器、一个开放的非平衡系统去看待。建筑作为一种物质组织，由组织中要素的秩序来控制空间中的能量流动，并以此平衡与维持组织的形式。物质、能量、气候、形式、身体和系统构成了热力学视角下建筑学体系的重要话语，对建筑中能量流动机理进行科学分析，可以为组织建筑本体形式、功能、空间等提供支撑。

建筑中的能量流动机理包括能量捕获、能量协同与能量引导等过程。“能量捕获”的关注对象包括风、光、太阳能及热能等；为了达到最佳的能量获得或抵制，建筑形态必须能良好地响应外部气候，成为空气、光线和热能的捕获器，以及一个多层次的热动力系统，从而建立环境要素和建筑形式之间的转换路径。“能量协同”是从看似混乱的建筑功能能耗中寻找逻辑关系并合理组织它们，通过立体配置的方式合理混合功能与面积分配，以实现建筑各部分之间的良好协作，平衡建筑内的能量流动，达到人体舒适与建筑低能耗的目标。“能量引导”则是对空气流动、光线导入与建筑之间能量流动的综合考量，能量可以在建筑内部或建筑之间流通和传递，建筑形成的能量引导通道成为空间组织的内在逻辑，并在建筑、建筑群体与城市等不同尺度上发挥作用。威廉·布雷厄姆将系统生态学家霍华德·奥德姆创立的能量系统语言引入建筑学，为建筑环境性能与能量利用评价提供了一个综合的、多尺度的图解（图 1）。系统语言通过相互嵌套的三种尺度与模式来阐述能量流动机理：建筑在城市环境中产生的对能量流动的强化作用；建筑作为隔离外部气候的庇护所所具有的热力学特征；建筑作为生活和工作场景对能量的需求。

能量在气候环境、建筑系统、人的身体之间流动与转化，这个热力学过程要求建筑成为气候与人的身体之间的热力学桥梁，建筑形式成为气候环境的转译与反馈。对光照、湿度、温度、空气流动等与能量梯度相关要素的关注与研究，能够促进与能量协同的建筑形式的生成。这是一种“能量形式化”的概念，在这个过程中如果将环境气候的力量最大化，就能启发建筑在空间、形态和性能上的突破，可称之为“形式追随能量”。如果考察一下丰富的前现代建筑遗产，就会发现在能量流动与建筑、城市形式之间存在丰富的关联。在也门希巴姆古城，夯土建筑构成的高层城市建筑聚落和迂回街巷形成了沙漠地区白天对于阳光的自遮蔽与夜晚对于热量的最大化储存，这是通过控制能量流动来调控建筑环境的经典案例（图 2）。

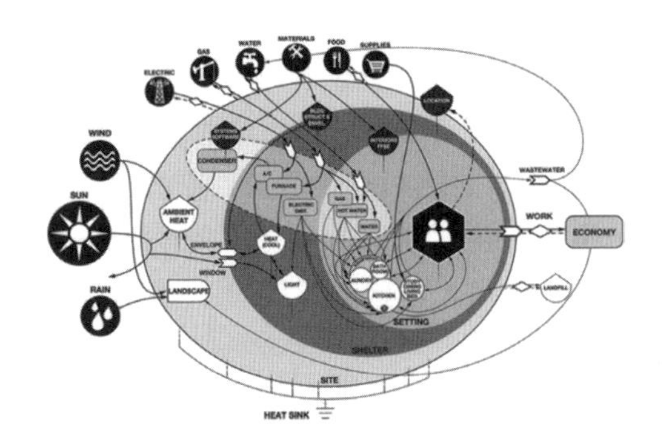


图 1 威廉 · 布雷厄姆的能量图解

图 2 也门希巴姆古城鸟瞰图

## 热力学建筑原型方法

“原型”一词自 2000 多年前的柏拉图始，最初多运用于哲学与宗教领域，柏拉图认为万事万物都有其被创始的原始模型，即若干个体之间享有的共同“理念”或“形式”。在建筑领域，维特鲁威认为建筑是对自然中的物体，例如对洞穴或燕巢的模仿；劳吉埃认为当今建筑来源于“原始茅屋”的原型。原型与初始的、根本的形态相关，是对复杂系统原初状态的追溯。对原型的研究涉及抽象思维、内在的法则以及挖掘形式和意象表层之下的结构。从原型到建筑，是一个转化、提升和植入的过程。“热源”与“热库”作为建筑中能量生产与能量消耗的两种模式，在建筑领域挖掘出洞穴与温室两种建筑原型。

热力学建筑原型将“形态生成学”引入设计，通过分析环境气候参数驱动，并将其可视化，揭示能量流动的热力学机理和性能，促使产生新的几何与形式，从信息数据、性能参数、能量流动到形式生成，在物质、形式、能量与性能这些热力学的核心话语间建立起一个全景视野。在这里，热力学法则成为塑造形式的首要工具，具有“原型话语”的潜力。热力学建筑试图重构建筑环境调控的主动式、被动式技术方法，并将其与建筑的建构形式紧密结合，尤其将被动式方法作为一个重要的关注点。2002 年，斯蒂芬 · 贝林著名的三角图解呈现了主动式、被动式与建构形式的关联，而伊纳吉 · 阿巴罗斯将其扩展成为过去（前现代）、现代主义与当代建筑的历史进化图解，从能量的视角展示了热力学建筑原型在环境调控方面的系统方法论。

针对热力学建筑的原型设计方法，一种整体的知识体系与观点至关重要。哈佛大学建筑与设计研究生院（下文简称“GSD”）的伊纳吉 · 阿巴罗斯、瑞纳塔 · 森克维奇与马提亚斯 · 舒勒等学者经教学研究，提出了“原型与协议”，通过对建筑中能量流动与系统特征的提炼，首先建立一种热力学的“定性知识”，基于气候数据的量化分析与研究，通过“列表与分级”，建立一套实用完整的气候、性能、材料的图解和图标；提炼出最具特征的环境要素和系统方法，在科学与文化的边界与冲突中，建立基本的建筑原型，并在形态上将其称为“热力学怪物”，是可与环境性能互动的，不以美学、建构学和现象学为首要生成动力的原型；进而通过植入城市环境的方式，检验和修正这些原型，提出城市建筑设计的新的路径与形式。

## 教学实验探索

基于与哈佛大学 GSD 合作的研究与教学，“热力学建筑原型”课程实验是针对同济大学建筑学专业四年级的设计课程教学——从 2014 年到 2018 年围绕“能量形式化与热力学建筑”主题展开的持续四年的教学计划，让学生通过一个学期的专门设计训练，初步探索自然系统与建筑本体设计互动的热力学方法。这一教学实验不是以建筑类型的划分为特征，而是聚焦于建筑与环境调控中的主导性要素与话题，课题设定的研究对象包括空气、光、热、自然系统、热力学建筑集群五个方面的专题。通过热力学原理与法则的研究，探寻建筑形式背后的隐藏逻辑，从而将热力学性能与建筑本体设计熔为一体，并探索基于中国城市与气候语境的热力学原型设计方法。

“热力学建筑原型”教学计划的设定包括八个主要步骤：案例研究提炼与模型还原、气候与自然特征数据分析、能量流动机理与系统模拟、热力学原型研究与优化、原型的建筑转化与实验检测、建筑的城市环境植入、热力学物质化与材料文化、研究与设计成果的整合。

在同济大学的教学实验中，热力学建筑的原型研究被扩展到“系统”的层面，而非仅仅局限于使用能量计算书等科学基础，两者的同步推进是原型设计的关键。这就从教学架构的基础层面避免了热力学计算书带来的原型局限性，是一种从系统组织与科学算法两个层面同时展开的原型探索。

教学包括五个重要的设计方法环节。第一个环节是精心挑选案例解析，其挑战在于“源于建筑，归于系统”。建筑被抽象为热力学视野下的系统组织，无论是岭南大屋的冷巷院落、蚁穴的通风系统，或是柯布西耶昌迪加尔的法院剖面，课程通过对案例的深入解读、抽象、提炼、模型表达与模拟验证，抽象出系统原理和关键词。第二个环节在于原型的生成，设计坚决摒弃“仿生”概念的形态或空间模拟，而指向一种结构化的系统抽象与解析。组织、架构、层级、多孔性、循环、流动、压力差、通道、烟囱效应、新陈代谢、过滤，这些是设计小组用来解释原型的关键词，原型的取舍不是一种形态、空间与美学的评判，而是一种具有转化潜力的能量流动与组织系统的评判。第三个环节是原型向建筑的转化，这是热力学原型设计最重要、最具挑战的部分，其中的根本在于原理法则的延续性与一致性，这种一致性通过环境分析的数据介入、气候图表的精准化模拟以及实物模型的检验得到保证，其中模拟风洞实验、光学实验等为设计原型的检测提供了扎实的技术支撑。在此环节，尺度的概念至关重要，原型逐步被赋予了建筑的尺度，建筑功能与空间的概念亦围绕热力学能量协同、能量捕获、能量引导展开，这有别于通常类型建筑设计中的功能排布、空间形态与流线组织。第四个重要环节是设计的城市环境植入与物质化表达。在此阶段，基于实际的城市基地与气候，建筑的布局、功能、形态、空间等本体设计成为关注重点，对原型的持续优化使之具备了物质化的条件，而热力学法则的延续、强化和细化则是设计的根本。最后一个环节是对“材料文化”的研究。通过对来自人类学的“材料文化”这一概念的借鉴，热力学建筑具有了文化意义的“在地性”，与真实的地域气候、通风、采光、热学需求以及热力学体验紧密相连。在此环节中，材料的热扩散率、多孔特征、组合特性、热力学性能是最为重要的关注点，同时，通过对材料与人关系的研究，舒适度与热力学体验最终介入了整个设计议程。

以上对设计进程进行了解析与探索，在环境的调控中，从对能量流动与数据驱动的气候分析，到原型生成和建筑转化等一系列环节，构成了热力学原型的完整设计流程，从而将“能量”作为一种结构化要素纳入建筑学的核心话语。

已经完成与正在进行的“热力学建筑原型”课程实验包括以下五个专题：

（1）风动：系列课程的第一次设计以“风”为要素，通过案例分析、专题研究、原型建立、软件模拟等抽象与具象的训练方式，对特定气候与风环境下的热力学原型加以探索，并以上海虹桥宾馆高层建筑的改造为载体。

（2）光：系列课程的第二次设计以“光”为主题，并以上海图书馆浦东分馆的专业竞赛课题作为载体，具有真实的城市环境与任务书要求。课程每个阶段有明确的成果要求，包括图纸与概念模型等；从课程初始的图书馆案例分析，到原型的研究与建立，再到材料实验与节点建构，对能量机制的关注贯穿始终。

（3）热塑形：系列课程的第三次设计以“热塑形”为主题，结合上海杨浦滨江电厂改造的实际课题与环境，课程既关注与体积、体量等相关的建筑外部属性，更关注与热力学相关的温度、压强、热容等建筑内部属性，热力学原型的研究也随之展开。

（4）自然系统：系列课程的第四次设计以“自然系统”为主题进行，更加强调建筑作为一个整体与自然的互动和响应，试图整合自然要素的综合效应，提出热力学建筑原型的设计策略。课题结合自然博物馆设计展开，并尝试提出中国北方气候特征下的建筑原型策略。

（5）热力学建筑集群：2014 年哈佛大学 GSD 与同济大学联合课程教学“热力学物质化——中国城市高铁站引导的热力学建筑集群”，以及 2015 年同济大学本科毕业课程设计“热带雨林——新加坡垂直城市集群”，均尝试了城市尺度上的热力学建筑原型，从而对课程实验进行了不同尺度的比较。

## 环境调控的形式法则

最后通过两个学生作业设计成果，我们可以检验一下在整体方法论基础上的形式可能性，以及设计提出的环境调控法则。

“塔风之下”提案源于中东地区的风塔——一种传统的被动式空气调节结构。首先对风塔模型进行了开口方向、开口大小等要素的变形，并置入风洞实验以研究入风角度与通风效能的关系；进而方案将风塔作为塔楼的原型进行扭转、拉伸、调节后植入场地，用一系列风环境模拟过程推敲原型在体量分割、三维肌理、当地气候、高度适应性与深度适应性上演变和优化的可能性；最终生成的方案形态带有明显的能量流动的态势，为“形式追随能量”提供了一个崭新的模式（图 3）。

“光合作用”提案从植物的呼吸作用与光合作用中寻找、捕捉、转化、传导、传输、存储和输出光能的机制，得出了一个连通着“热力学烟囱”的图书馆原型：烟囱底部作为一个个漫游式阅读空间，部分作为城市公园对公众开放；上层与烟囱中下部脱开并拥有稳定的风环境与热环境；而中间层则随着季节、时间的变化有多样的风和光。提案以孔隙大小、烟囱通道形态与不同层的厚度等参数作为原型的变量，并将得出的各个原型剖面模型置入光环境实验，研究其分别在春分、夏至、冬至一整天的自然采光性能。选择并优化后的原型则以不同尺度赋予其不同功能，组合和植入环境而得到建筑的基本形态。最终，

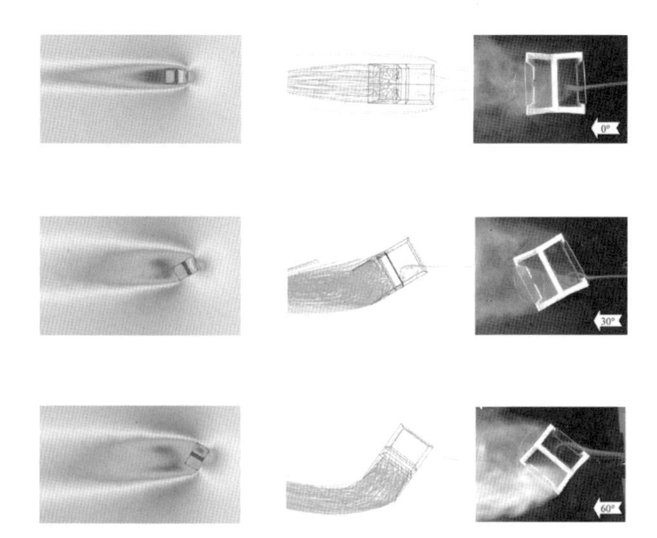


图 3 “塔风之下”提案：风环境模拟

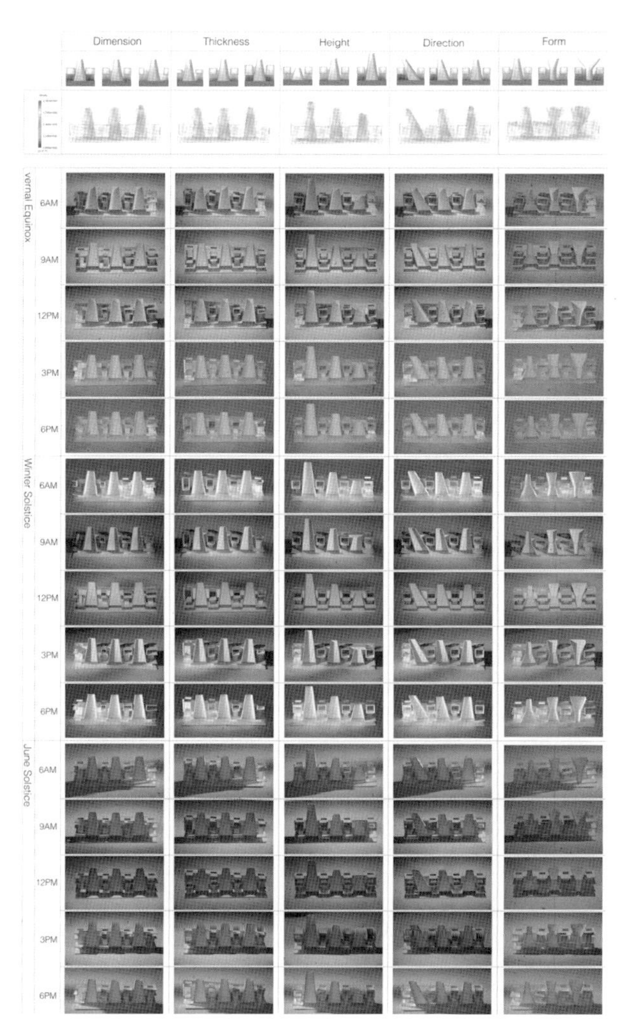


图 4 “光合作用”提案：各个原型在不同季节的采光模拟

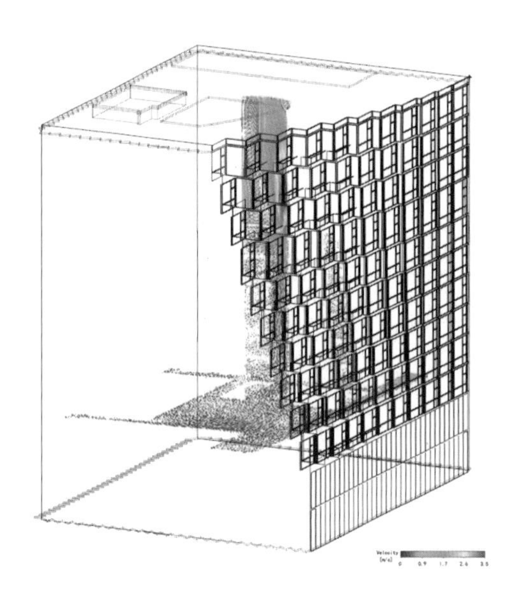

将得出的形态进行建筑学意义上的深入发展，如采光井尺度的调整和风塔倾角的变化。此外，提案也对建筑界面的材料与构造做了更进一步研究，将表面的开孔分为里外两层并轻微转动，尝试通过控制开孔大小和转动角度来控制不同功能区域的光照条件，同时也能增强烟囱的拔风效应。该组同学的中期成果获得上海图书馆东馆专业组竞赛二等奖（图 4）。

## 建筑设计的方法论革新

热力学建筑原型源于对建筑环境调控的历史性与理论性思考，以技术文化的观念重构建筑思考的整体性。相关的系列教学研究通过对环境要素的引入和设计参数的整合，研究能量在建筑中的生产、流动和消耗过程，思考建筑作为一个热力学系统，其空间、结构和组织等的原型潜力和设计转化。对与课程教学同步的建造实践的考察与关注，也成为教学非常有力的支撑（图 5）。

伊纳吉 · 阿巴罗斯认为，对能量形式化和热力学建筑的关注代表了当今设计文化认识论转变的最激进和最重要的方向之一：对于能量与资源可持续的关注不应成为形式的制约，也不是与理论设计方法论无关的技术标准；它是带有“范式转变”潜力的，意味着建筑设计方法的转折。作为设计教学与研究的一个阶段性总结，我们期待借此在建筑设计方法论的革新方面做出持续性探索。

图 5  崇明体育训练中心（主持建筑师：李麟学）

**本文由同济大学“高密度人居环境生态与节能教育部重点实验室”联合研究中心资助。周渐佳博士共同参与了课程建设与教学工作，伊纳吉 · 阿巴罗斯教授深度参与了课程指导，陶思旻、何美婷、侯苗苗、李鹜、吕悠、烘烽桓、葛康宁、苏家慧、郭绵沅津等参与了助教工作。**

参考文献

[1] William W. Braham, Daniel Willis. Architecture and Energy: Performance and Style[M]. London & New York: Routledge, 2013: 3.
[2] Reyner Banham. Architecture of the Well-Tempered Environment[M]. University Of Chicago Press, 1984.
[3] Kiel Moe. Building as Refrigerators [J]. Solid Harvard GSD Series a+t, 2015,(Interior Matters): 20-28.
[4] Bruno Latour (Author), Catherine Porter (Translator). We Have Never Been Modern[M]. Harvard University Press, 1993.
[5] Howard T. Odum. Environment, Power, and Society for the Twenty-First Century: The Hierarchy of Energy[M]. Columbia University Press, 2007.
[6] Adrian Bejan, J. Peder Zane. Design in Nature: How the Constructal Law Governs Evolution in Biology, Physics, Technology, and Social Organization[M]. Doubleday, 2012.
[7] D'Arcy Wentworth Thompson. On Growth and Form[M]. Create Space Independent Publishing Platform, 2011.
[8] Kiel Moe. Insulating Modernism: Isolated and Non-isolated Thermodynamics in Architecture[M].Basel: Birkhäuser, 2014.
[9] William W. Braham. Architecture and Ecology Systems-Thermodynamic principles of environmental building design in three parts[M]. Routledge, 2016.
[10] Kiel Moe. Convergence: An Architectural Agenda for Energy[M]. Routledge，2013.
[11] Victor Olgyay. Design With Climate: Bioclimatic Approach to Architectural Regionalism[M]. John Wiley & Sons, 1992.
[12] Sanford Kwinter. Far from Equilibrium[M]. Barcelona/New York: Actar, 2007.
[13] Iñaki Ábalos. Prototypes and Protocols[M]. AV Monographs 169, 2014.
[14] Iñaki Ábalos. Why we don't draw arrows (hardly ever)[M]. AV Monographs 169, 2014.
[15] 李麟学. 知识 · 话语 · 范式——能量与热力学建筑的历史图景及当代前沿[J]. 时代建筑，2015(2)：10-16.
[16] Yasha J. Grobman, Eran Neuman. Performalism: Form and Performance in Digital Architecture[M]. Oxon: Routledge, 2012.
[17] 李麟学，叶心成，王轶群. 环境智能建筑[J]. 时代建筑, 2018（1）：56-61.
[18] Scott Marble. Digital Workflows in Architecture[M]. Birkhaeuser, 2012.
[19] Rashida Ng, Sneha Patel. Performative Materials in Architecture and Design[M]. Chicago: Intellect, 2013.

# Thermodynamic Archaeology: Innovation of Architectural Design Methodology

**LI Linxue**

A background that energy and thermodynamic architectures are valued in contemporary times is the inherent crisis of architecture in an extensive context of globalization and sustainability. As said by Sarah M. Whiting, an American architectural theorist, "the traditional architecture and the autonomy of its theories are being challenged as the contemporary architecture increasingly focuses on issues such as ecological balance and social responsibility." While we refocus on the energy issue and thermodynamics inside a building, it is necessary to involve archaeological survey and reconstruction of a knowledge system. When the laws of thermodynamics are applied in the field of architecture, "air" becomes the protagonist of space arrangement; a building can be understood as a way to arrange different materials, and this arrangement brings order to "energy flow" while balancing and maintaining the "form of matter" of architecture.

A series of seminars and design studios regarding thermodynamic architecture have been held at College of Architecture and Urban Planning of Tongji University since 2015. Moreover, the design studio of "Thermodynamic Architectural Prototype," a full-semester-long elective course for senior undergraduate students, has become a platform for the compact integration of theoretical research and the pedagogy of design. In the original course framework, knowledge production and design exploration became the two pivots of this studio. The systematic archaeological study of the knowledge in energy and thermodynamics and the inclusion of the interdisciplinary knowledge in nature, systematics, cybernetics, ecology, meteorology, and thermodynamics have become indispensable parts to the design studio. The formal design and practical operation are deeply built upon and supported by the production of knowledge on a broader background. As pointed out by Kiel Moe: "Finally, the actual thermodynamics of building, rather than the violent abstractions of energy analysis and optimization, are understood as fundamental to an alternative, more cosmopolitan—in the most literal sense of the term—politics for built environments."

### Environmental Control and Architecture as Open System

In the book *The Architecture of the Well-tempered Environment*, Reyner Banham explains the issues of environmental control and design strategies faced by the architects in the era of mechanical industry. While modern architecture relies on technical equipment, he mentions, "Under the right circumstances, a system that can subtly deal with the interrelation between human and environment does not necessarily lead to a complex mechanization." As his comment on Wright's design of Baker's house, Wright's consideration of machinery and equipment is not only to install them properly, more importantly, is to allow them cooperating with building structures, functions, etc., so that the overall performance is better than that of part-to-part.

Kiel Moe, who is considered as the Banham in the 21st century by Iñaki Ábalos, holds a critical point of view toward environmental control since modern architecture. In his book *Insulating Modernism: Isolated and Non-isolated Thermodynamics in Architecture*, he reflects the convention of building mediation and insulation since modernism, pointing out that the technical system diverges from the "Principles of Thermodynamics," so he calls for re-evaluating architecture's "isolated system". The isolation is characterized by the "man-made weather" brought by air-conditioning and ever-increasing structural insulation of the buildings, which is called "architecture as refrigerator" by Kiel Moe. The pre-modern "climatic architecture", which had spread around the world before the popularization of air conditioners in the 1940s, was replaced by "hermetic architecture" of modernism. However, the so-called energy-saving buildings certified by LEED, in some way, are still the extension and intensification of the "environmental isolation." This kind of environmental isolation has become one of the causes of massive energy consumption and environmental crisis. At the same time, due to air-conditioning, the homogenization of interior spaces has led to what Koolhaas called, the "junkspace." The construction logic of the building itself also often becomes the victim of an equipment upgrade. At a macro level in urban and ecology, the intensification of "internal efficiency" brought by enclosed building system is at the expense of higher energy consumption in the entire life cycle and loss of urban space vitality.

According to Ilya Prigogine's second law of thermodynamics, architecture is a "non-equilibrium open system" and a thermodynamic "dissipative structure." This dissipative body is characterized by the maximization of energy exchange and maintenance of entropy which must be considered in a holistic system of thermodynamics. Introducing the knowledge system of thermodynamics into architecture field has brought about a new, archeological architectural environment mediation paradigm. In Adrian Bejan's system diagram, the open system is a fundamental structural shift. Thermodynamic architecture is not an "additional" manifestation of architecture, nor does it simply aims at energy efficiency and energy conservation. It is based on the "structural" goal. As Deleuze calls, "dynamics generates form," it takes material energy as a source of design dynamics and explores the design potential of architecture as an open system. Through in-depth exploration of energy, material, and form, thermodynamics can touch on the "autonomy" of the building and possess the "instrumentality" of environmental performance, in the end, achieve an "autonomous instrumentality." In this sense, thermodynamic architecture attempts to construct a formal principle that provides a new paradigm for building environment control, thereby the idea of sustainability from the perspective of operating principles of energy can be brought into architecture field. This is not only related to the technical level, but also a critical, cultural, and political architectural agenda.

**Energy-flow Mechanism and Diagram**

We take energy as the focus point of architecture research again and apply thermodynamic methodologies in the studies of architecture, which means considering architecture as a thermodynamic container of energy and an open, non-equilibrium system. As a kind of material organization, architecture controls the flow of energy in space by the element order in the organization, to balance and maintain the form of the organization. Material, energy, climate, form, body, and system constitute the essential discourse of architectural study under thermodynamics. Through analyzing the mechanism of energy flow in architecture scientifically, it is possible to provide support for the organizational structure, function, and space.

Architecture's energy-flow mechanism includes energy capture, energy coordination, energy channel and other processes. "Energy capture" focuses on wind, light, solar energy, and thermal energy. To achieve the best result of energy gain or resistance, the architectural form must respond properly to external climate, and become a catcher for air, light, and heat, as well as a multi-layered thermodynamic system that can establish transitional path between environmental elements and architectural forms. "Energy coordination" seeks logical relationships from chaotic energy consumption of building functions and organizes them in a rational way. Through three-dimensional configuration, mixed function and area distribution, it can achieve a good collaboration between various parts of the building and balance the energy flow of the building, to meet the needs of comfort and low energy consumption. "Energy guidance" is a comprehensive consideration of air flow, light introduction, and energy flow between buildings. Energy can be circulated and transmitted within or between buildings. The energy-guided channels formed by buildings become the internal logic of spatial organization which plays a role in various scales, such as buildings, building clusters, and cities. The language of energy system conceived by an ecosystem ecologist, Howard T. Odum, is introduced into architecture by William W. Braham, which provides an integrated, multi-dimensional diagram for assessing the performance of architectural environment and energy consumption (Fig. 1). The system language elucidates the energy flow mechanism through three scales and patterns nested within each other: architecture's intensifying effect on the energy flow generated in an urban environment; its thermodynamic characteristics as a shelter from external climate; architecture's demand for energy in living and working scenario.

Energy flows and transforms between climate, architectural system, and human body. This thermodynamic process requires architecture to become a thermodynamic bridge between climate and the human body while architectural form becomes the translation and feedback of climatic environment. Through the attention and research on the elements related to energy gradients such as light, humidity, temperature, air flow, etc., it forms a type of architecture that can advance energy coordination. This is a concept called "formalization of energy." During this process, it tries to maximize the power of environmental climate, thereby stimulating architecture's possibility in space, form, and performance. In other words, "form follows energy." If one looks at the rich heritage of pre-modern architecture, there is a profound correlation between energy flow and architectural and urban form. In the ancient city of Shribam in Yemen, high-rise urban settlements formed by rammed earth structures and winding streets and lanes help the building shield from sunlight in the desert during the day and maximize heat storage at night, which is a classic case of managing building environment by controlling energy flow (Fig. 2).

**Methodology of Thermodynamic Architectural Prototype**

The term "prototype" can be traced back to 2000 years ago in Plato's time, When it was used in philosophy and religious studies. Plato believed that though everything had its original mode when created, there existed the common "ideas" or "forms" shared by a group of individuals. In the field of architecture, Vitruvius believed that architecture was an imitation of objects in nature, such as caves or sheds made of trees. Similarly, Laugier believed today's architecture was derived from the prototype of "primitive hut". "Prototype" is related to the initial, fundamental form and is a resource to the original state of a complex system. The study of "prototype" involves abstract thinking, internal principles, and revealing the structures hidden behind form and imagery. From prototype to architecture, it is a process of transformation, elevation, and implantation. "Sources & Sinks" are used as two models of energy production and consumption in architecture and are developed into two kinds of architectural prototypes—cave and greenhouse.

By means of analyzing environmental climate parameters, thermodynamic architectural prototype introduces the idea of "morphogenesis" into the design and visualizes it, so it can reveal the thermodynamic mechanisms and performance of energy flow, and stimulate new geometries and forms from information data, performance parameters, the energy flow to form generation. In this way, we can create a panoramic view among matter, form, energy, performance, and other thermodynamic keywords. Here, the laws of thermodynamics with the potential of "prototype discourse" become the primary tool for shaping forms. Thermodynamic architecture attempts to reconstruct the active and passive technological methods of environment control, and integrate it closely with constructive forms, especially focusing on the passive part. In 2002, Stefan Behling's famous triangle diagram presented the connection between active, passive, and constructive forms. Subsequently, Iñaki Ábalos expanded it into the historical evolutionary diagram of the past (pre-modern), modernism, and contemporary architecture. From the perspective of energy, it illustrates the systematic methodology of environmental control in thermodynamic architectural prototypes.

It is essential to see the prototype design method of thermodynamic architecture in a holistic knowledge system and viewpoint. Therefore, Iñaki Ábalos, Renata Sentikiwicz, and Matias Chuller of Harvard Graduate School of Design ("GSD") proposed the research of "prototypes and protocols." Through the extraction of energy flows and systematic characteristics in architecture, their first step is to establish a "qualitative knowledge" of thermodynamics. Based on quantitative analysis and research of climatic data, through the method of "listing and ranking," they then establish a set of practical and complete illustrations and diagrams of climate, performance, and materials. Then they extract the most distinctive environmental elements and systematic methods. Along the boundaries and conflicts between science and culture, a basic architectural prototype named "Thermodynamic Monster" is established. This is a prototype that interacts with environmental performance without using aesthetics, construction, and phenomenology as the primary dynamics for a generation. At last, these archetypes are test and corrected by being implanted into the urban environment. Through the examination and revision of these prototypes, it is possible to emerge new ways and forms of urban architecture design.

**Teaching Experiments**

The course "Thermodynamic Architectural {rototype" is a collaborative effort with Harvard University Graduate School of Design. It is specially designed for senior students majored design studio at the School of Architecture in Tongji University. From 2014 to 2018, the teaching plan is designed to be a one-semester-long training of design exercises and mainly conducted on the subject of "energy formation and thermodynamic architecture," which aims at exploring thermodynamics strategies that can interact natural systems with architecture. This teaching experiment does not focus on dividing architectural typology but on the dominant natural elements and topics in architecture and environment control. The subjects selected for the course include air, light, heat, natural systems, and thermodynamic architecture clusters. Through the study of thermodynamic principles and laws, students can explore the hidden logic behind architectural forms, which helps to integrate thermodynamic performance and architectural design. Finally, students can explore the typical thermodynamic prototyping methods within Chinese urban and climatic contexts.

The teaching plan of the "Thermodynamic Architectural Prototype" course is divided into the following eight steps: case study and model reduction; data analysis of climatic and natural features; energy-flow mechanism and system simulation; thermodynamic prototype research and optimization; architectural transformation and experimental testing of prototype; architectural implantation in urban environment; materialization and material culture of thermodynamics; integration of research and design results.

In the teaching experiment of Tongji University, the prototype research of thermodynamic architecture is extended to the "system" level, without limiting to the energy calculation formulas under scientific basis. The key of the prototype design is to integrate both aspects, which fundamentally avoids the limitations of prototypes due to thermodynamic formulas. It is a prototype exploration in terms of system organization and of scientific algorithms. The course essentially involves five design components. The first part is the selection of case studies, whose challenge is that it "is derived from architecture while still a part of the system." Architecture is abstracted as a system organization in terms of thermodynamics, be it the cold alleys in Lingnan, the ventilation structure of the ant hole, or the section of Corbusier's courthouse design in Chandigarh. Through in-depth studies of the case, through abstraction, refinement, modeling, and simulation, the abstract system principles and keywords become the point. The second important component in teaching lies in the generation of prototypes. The design resolutely abandons the "bionics" concept in terms of morphology or spatial simulation. Instead, it points to a structured system abstraction and analysis. Organization, architecture, hierarchy, porosity, circulation, flow, pressure difference, channel, chimney effect, metabolism, and filtration are some of the keywords used by the design team to articulate "prototype." The choice of a prototype is not an assessment of form, space, or aesthetics, but the energy flow and assessment of organizational structure with a potential of transformation. The third and the most important component is the transformation to architecture, which is the most challenging part of thermodynamic prototype design. It is the continuation and consistency of the principles and laws. By means of plugging in the data of environmental analyses, precise simulation of the climatic diagram, and validation of the physical model, the "consistency" is ensured. Meanwhile, simulated wind tunnel experiments and optical experiments provide solid technical support for the testing of design prototypes. In this component, the concept of scale is of utmost importance. Prototypes are gradually given the scale of architecture. The concept of building function and space is also centered around energy coordination, energy capture, and energy guidance in terms of thermodynamics. This is different from the functional programming, spatial form and circulation organization in architectural design. The fourth important component is the implantation into urban environment and materialization of the design. At this stage, the actual urban site and climate, architectural planning, function, form, and space become the focus of attention. The continuous optimization of the prototype makes it materialized, and of course, the continuation, reinforcement, and refinement of the laws of thermodynamics are fundamental to the design. The last component of the course is the study of "material culture." Through reference to this concept borrowed from anthropology, thermodynamic architecture possesses the notion of "site-specificity" in a cultural sense, which is closely interconnected with the actual geographical climate, ventilation, daylighting, thermal requirements, and thermodynamic experience. In this component, the thermal diffusivity, porous characteristics, combined properties, and thermodynamic properties of the material are the most important concerns. At the same time, through the study of the relationship between materials and human, comfort and thermodynamic experience are eventually involved in the overall design agenda.

Through the above analysis and exploration of the design process, in the environmental mediation, through a series of components from energy flow and data-driven climate analyses, to prototype generation and architectural transformation, a complete design flow of the thermodynamic prototype is established. Thus "energy" as a structural element is incorporated in the core discourse of architecture.

The following are five topics in the course (some are completed while some of them are still in progress).

(1) Air in motion: The first design exercise of the course takes "wind" as the key factor. Through adopting abstract and concrete training methods such as case analysis, specific research, prototype building and software simulation, it explores the thermodynamic prototype under specific climate and wind conditions, and uses the reconstruction of high-rise buildings of Shanghai Hongqiao Hotel as the medium of the exercise .

(2) Enlightenment: The second design exercise of the course takes "light" as its theme and takes the professional competition for Shanghai Library Pudong Branch as the medium. It deals with an actual urban environment and needs to meet the requirements outlined in the design brief. Each stage of the exercise requires a concrete delivery, including drawings and conceptual models. From the initial case studies on the library, the research and establishment of prototypes, to the material experiments and construction, the focus on energy mechanism is attended throughout the design process.

(3) Heat formation: This exercise combines with the practical exercise and environment of restructuring the Yangpu Riverside Power Plant in Shanghai. It tackles architecture's external properties such as mass and volume, more so with the internal properties that are related to thermodynamics such as temperature, pressure, heat capacity, etc.. The study of thermodynamic prototypes is integrated into the process.

(4) Natural system: The fourth design exercise emphasizes the interaction and response of architecture as a whole with nature,

trying to integrate the comprehensive effects of natural elements, and proposing prototype design strategy of thermodynamic architecture. The exercise is developed in conjunction with the design of Natural Museum and attempts to propose an architectural prototype strategy that meets the climate characteristics in northern China.

(5) Thermodynamic conglomerates: In 2014, Harvard's GSD and Tongji University jointly taught the course "Thermodynamic materialization—thermodynamic building cluster guided by China's urban high-speed rail station," and in 2015, Tongji University held a undergraduate program "Tropical rainforest—Singapore's vertical city clusters." Both studios tried prototypes of thermodynamic architecture on an urban scale, and compared the experiments to various scales.

**Formal Principles of Environmental Mediation**

Now, through two student assignments, we can examine the formal possibilities based on the holistic methodology and the principles of environmental mediation proposed by the design.

The proposal "Under the wind" takes inspiration from the wind tower in the Middle East—a traditional passive air-conditioning structure. It alters a few structural factors of the wind tower such as the opening direction and the size of the opening, and puts them into the wind tunnel experiment to study the relationship between air inlet angle and ventilation efficiency. The program then uses the wind tower as the prototype of the tower after performing torsion, stretching, and adjustment on it and implants it into the site. A series of wind environment simulations are conducted to deduce the prototype in volume segmentation, three-dimensional texture, local climate, the possibility of high adaptability and depth adaptability in terms of development and optimization. The resulting form of the program has a clear flow of energy, providing a new model for "form follows energy" (Fig. 3).

The proposal "Photosynthesis" derives its mechanism of locating, capturing, transforming, transmitting, transferring, storing, and exporting light energy from the plant's respiration and photosynthesis, and comes up with a prototype of a library connected to a "thermos-chimney": the bottom of the chimney is used as a roaming, reading space, partially opened as a urban park to the public; the upper part of the chimney is separated from its middle and bottom parts, with a stable wind and thermal environment; the wind and light condition in the middle part varies as the season and time changes. The proposal adopts parameters such as pore size, the shape of chimney channels, and thicknesses of different layers as the prototype's variables, and puts the section model of each prototype into light environmen experimentation to study the performance of natural daylight throughout the spring equinox, the summer solstice and the winter solstice. The selected and optimized prototype is then given different functions based on different scales, and a basic architectural form is generated after the prototype is put into the environment. In the end, the resulting form is further developed in the architectural sense, such as the adjustment of the lighting well scale and the change of the wind tower inclination. The proposal also studied the material and structure of the architectural interface, dividing the surface opening into two layers, the inside and the outside, which can be turned slightly to adjust the sizes and the rotation angles of the opening in order to control lighting conditions for different areas. At the same time, it can also enhance the chimney's pull-out effect. The middle-term result of this group of students won the second prize of the professional competition for Shanghai Library East Branch (Fig. 4).

**Innovation of Architectural Design Methodology**

The prototype of thermodynamic architecture derives from historical and theoretical considerations of the mediation of the built environment, reconstructing the integrity of architectural thinking from the perspective of technological culture. Through the introduction of environmental factors and the integration of design parameter, this series of research-oriented teaching aims to study the process of energy production, flow and consumption in architecture, explore the prototyping potential and design transformation of space, structure, and organization by treating architecture as a thermodynamic system. A look into some practices regarding thermodynamics also helps the teaching program (Fig. 5).

Iñaki Ábalos believes that the concentration to energy formation and thermodynamic architecture represents one of the most radical and important directions in the shift of today's design culture epistemology: the sustainability concern about energy and resources should not restrain the form of an architecture, nor should it be a technical standard that is unrelated to the theoretical design methodology; instead, it is a potential with a "paradigm shift" and a turning point in the methodology of architectural design. Here, with this stage summary on our teaching program, we look forward to further exploring the innovation of architectural design methodology.

**This article was funded by the Key Laboratory of Ecology and Energy-saving Study of Dense Habitat at Tongji University. Dr. Zhou Jinjia participated in the design and teaching of the course. Prof. Iñaki Ábalos played an advisory role for the course. Teaching assistants include Tao Simin, He Meiting, Hou Miaomiao, Lu You, Bakhan, Ge Kangning, Su Jiahui, and Guo Mianyuanjin.**

# 为何我们不用箭头？

## 伊纳吉·阿巴罗斯

人们有时会问：热力学这一概念是如何运用到设计过程中的？为什么我们不愿意展示设计过程“后厨”的玄机？人们期待的答案或许是带箭头的图解、红蓝相间的分级图示，或者缺乏创造的“环保技术”图表。

无论从学术还是实践角度出发，我们都不希望使用、展示这些工具，而是期望通过在建筑学以及在几乎所有历史建筑中（除了现代建筑）都存在的整体观，做到定性知识先行。这些原则如此重要，现如今却几乎“不复存在”。

我们的出发点是将热力学原理以定性的方式融入设计过程，这是建筑师的主要工作内容。定量分析已经进入工程领域，尽管它们彼此间边界模糊、渗透积极（尤其是在复杂的环境中），有必要说明的是：定性近似不构成图表，且无法通过计算机程序学习。定量分析只是基于个别原则和一系列公式，并且在实际的天气信息或材料表（焓湿图、阿什比材料剖面图等）中都会加以简化，这些表格不需要展示，但确实有用。其他的概念性文件，如贝林的三角形，则准确地体现出需要优先考虑的与形式、物质相关的问题，从开始就实现建筑良好的热动力性能，这对避免之后额外的工作很重要（传统设计过程总是会出现的情况），不会造成返工或昂贵的维护及施工。

我们向学生建议的方案很好地阐述了我们在国际专家的协助下逐渐实现个性化的过程（参见《原型与协议》一文）。一旦投入使用，很难脱离这些流程，因此我们称之为“热力学怪物”，让人上瘾（能识别项目中可能有效的一切）。热力学和建筑是综合的，就像工程总是根据“列表与排名”这一双重过程将关键要素排在最优先级。

分析各地气候的积极因素，深入研究气候最具特色的方面（有时很容易忽略，如马德里的夏夜微风，其影响在图表上可能并不显著，但对于夏夜的“热动力重启”至关重要），这如同从不推延的庄重仪式，先于草图或图纸的探讨，对于结构、形式和物质的考量十分必要。

最初的视角有时会带领我们进入未曾预见的领域。比如，多次分析马德里的气候后，一项更多关注其对全球重要性的新研究表明，马德里是唯一可以重新考虑其住房和办公室市场为100%被动式的欧洲大都市，这也预示着一项改造翻新历史名城的政治议题，而这是量化研究无法企及的。因此，我们与马提亚斯·舒勒在哈佛设立了图形化系统设计课程，舒勒的实验性内容也显示出这一假说的巨大潜力，重新评估传统的介入技术和方法，改善对健康和舒适度的影响（大幅减少热岛效应和环境污染），以其他方式促进可持续发展，如今还开启了专业文化与当代科学之间的关系。这些目标显然与其他科技至上论截然不同。

我们发现建筑能源认证存在一些矛盾之处，这些矛盾提醒我们不要掉进因为商业性而过分简化的陷阱，比如使用被动施工方法无法得到LEED证书，因为不用电（空调），能效无法测量。这种矛盾在评级系统中对公共空间的作用更加明显。基本上建筑能源证书仅涉及内部效率，并未提及热能增益的去向——全部释放给了公共空间。建筑围护效率越高，街道越有可能成为热能废物场，特别是轻质、后部通风的立面。我们只有通过研究建筑的热质量（及其立面内和/或外的位置）对公共空间舒适度的影响，才能考

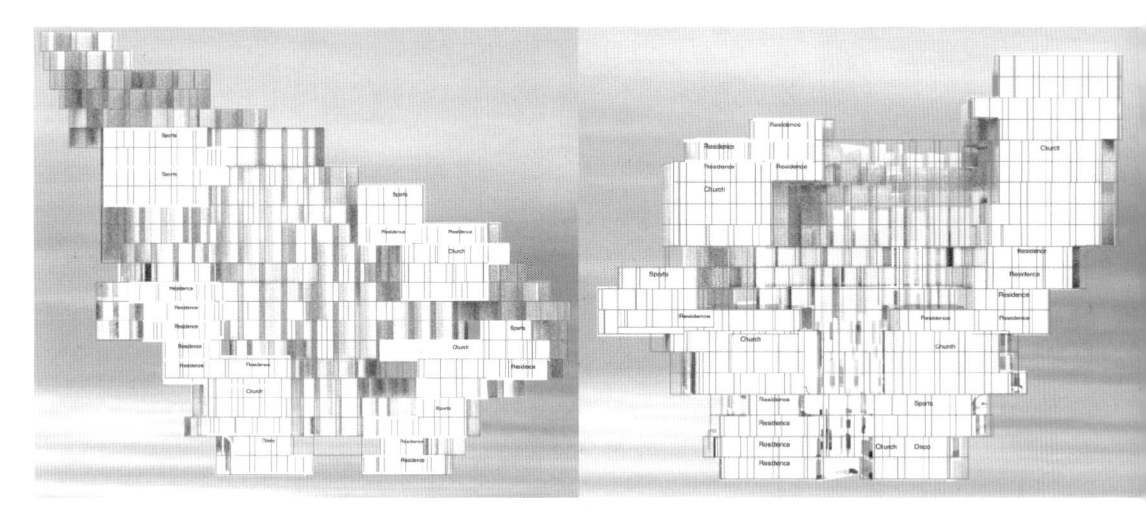

“热力学怪物”的聚集，HUANG Xiaokai、SU Ke

虑建立一个平衡公共和私人领域的体系：二者的边界绝非相互抵消，而是有某种走向。另一个例子，我们无法评估城市树木的积极作用（比如测量树荫这一最廉价高效的操作系统的作用，更不要说其美观和生物多样性，以及对地中海和热带城市舒适度的改善），因此量化数据将之排除于任何体系之外。尽管现在这些方面有被纳入的趋势，但设计过程对它们的考虑还是少之又少。

## 创意热力学

毫无疑问，当气候条件颇具挑战性（如湿度高、气温高）或者项目条件使设计难以展开，抑或是当我们更自由地去尝试新想法时，具有科学背景的专业人士的帮助是不可或缺的，并且在项目之初就应当予以考虑，尤其是在材料科学及建筑物理学领域。这有助于以综合的视角分析热力学的作用，从全球的角度考虑我们应该采取怎样的系统。威利斯·开利发明的空调系统将气流像水一样处理，提供冷气和暖气的方式，非常低效且不利于健康，并且导致建筑内气流量最小化，经济性差，而建筑也只能是绝缘封闭的场所。

基尔·莫准确地指出，把建筑变成了冰箱着实是残忍的，消除了文化、环境和社会的互动，因为它所依据的原则与城市概念截然相反，旨在尽可能抵消所有与外部的交流。

热质量已经被热绝缘取代；自然或对流通风被建筑密封所取代。材料唯一的活跃参数是传导性；对热舒适性十分重要的辐射因素和扩散率等因素却常常被忽视。因为一直以来，热分析的方式是静止的，忽略了气流分布的时间和变量，而日常循环不断变化，热液的变化对舒适度至关重要。这不是简单地质疑这些系统的重要性，而是要理解技术和文化条件对克服问题、创造原型、建造形式和物质构成产生的建筑综合体的意义，而这一点对于地球上几乎所有的建筑并不新鲜。然而现在却鲜有这样的例子了，我们面临的情况变了，规模、功能、密度、与基础设施的关系、全球化的物质文化，这一切使我们无法回到过去。当代建筑师就像孤儿，历史的孤儿以及当代先驱的孤儿，不得不遵从基于分析系统的规范，而规

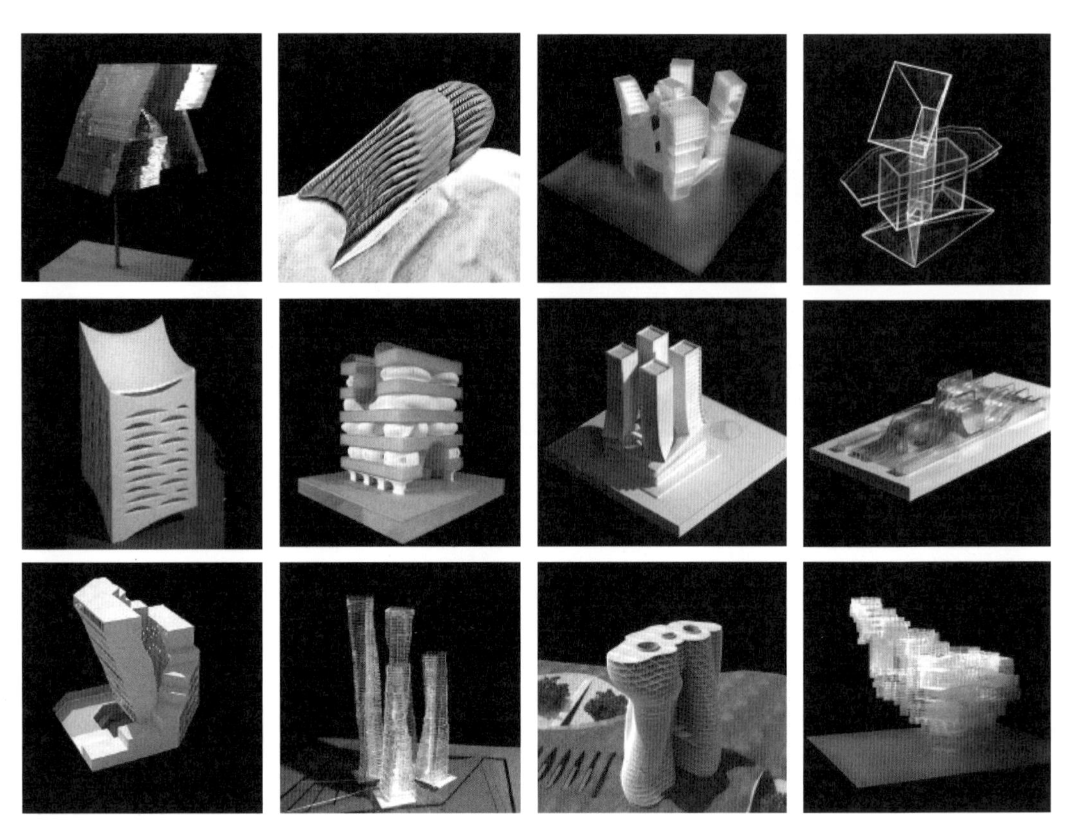

哈佛大学设计研究生院学生模型

关于空间组织、材料与室外舒适度参数工具之间相互作用的研究，Felipe Silva、Ziyin Zhou

范歪曲了现实。我们应该理解资源并且以独特的方式加以利用，现在也有足够的数据帮助我们趋利避害。

大面积使用数字化工具推动了室内空间（家庭或办公）的机械化，我们可以重新考虑室内热压力的概念。孩子的娱乐室布满了数字设备，犹如小型数据中心或高科技办公室。住户与电器设备的合力使生活方方面面的用电量攀升，因此亟须通过能源交换循环加以管理及调整，有时这与室外日常或季节性交换循环相关联。在高密度城市，混合功能建筑不仅是好的商业选择，还可以打造一个开放的系统和耗散结构，这样的系统和结构在不久之前还仅适用于生物或生态环境，而非建筑。

## 建筑的调解

我们可以继续列举建筑及城市热动力学相关的定性概念带来的相关事宜，不仅仅是定量方面，还有定性方面（即创意）。回到本文的主旨，我们理解这一认识是关键的文化和政治武器，同时在美学方面不乏创意，这就是建筑领域的热力学法则最有趣的角度。综合性视角既关注理论，又退后一步纵观全局。总而言之，我们对过程伊始尤为关注，确保选择方案之前，做好第一个决策：针对形式、质量、气流、与自然元素的关系、管理室内外的热收益与经济上的合理性，实现项目的良好开端。我们会格外质疑适用于公共领域、源自充满漏洞的传统方法的形式和施工假说，因为其中包含的策略会影响生活的各个领域。我们尤为关注广阔且具有创意的全景，当在实践及理论层面重新从热力学的角度考量城市时，这一全景就展开了：为什么斟酌这一看似微不足道的因素，可以重新考虑那些我们曾经毫不质疑的规模、方法及目标呢？这适用于学术领域以及专业领域（不同程度的完美）。事实上，由此构成的反馈系统至关重要，重新界定了学术作用，以及建筑在科学、文化和社会等方面的调解作用。就像有些景观学教授禁止学生将“绿色”作为优先策略，以避免落入俗套一样，在我们的办公室或课堂也很少见到箭头图示，除非是在描述复杂的现象时，我们将无形的动态现象视觉化的能力是有限的。由于形式与物质结合，与构成动态气候环境的自然元素互动，热力学法则在不同的审视方式下，或隐蔽或明显。做完相关的定量分析后，不再进行调整，这些材料组织才能促进出现热力学行为。

然而这些材料组织本身就是建筑，通过简单的公式和原则加以解释，而我们工作的目标，尤其是涉及学术项目时，是推动打造突破体验局限的建筑，并为 21 世纪的城市遇到的困难提供解决方案。

（请不要误解：这些问题绝非质疑定量知识的价值，而是启发思考其他要求，这比考虑增加两厘米的隔热层是否意味着认证评级的标准由铜级改为银级更加重要，虽然后者是所有建筑领域环境计算程序以及与此紧密相关的能源认证图章的焦点。）

# Why We Don't Draw Arrows?

Iñaki Ábalos

Sometimes they ask us how the concept of thermodynamics is brought into our design process and why we are so reluctant to show the operational "kitchen" of these processes, probably because the sustainable cliché seems to ask for diagrams with arrows, sections with "Illustrator" gradations from red to blue, or multicolor "Ecotec" graphics.

What we are trying to accomplish both in the academic field and our professional practice involves neither using nor displaying this paraphernalia, but prioritizing qualitative knowledge through architecture and a holistic view that has been present in almost all historical architectural styles (but only exceptionally in the case of modern architecture), and which is almost a "dead tongue" today; though it is probably more necessary than ever.

We start from the basis that a qualitative integration of thermodynamic principles in the design process is the architect's main working matter. Quantitative analyses enter the sphere of engineering and, though the frontiers are flexible and permeability is positive especially in complex contexts, it is necessary to affirm that qualitative approximations are not graphic material, and can't be learned using a computer program. They are based on few principles and a series of formulas, and are simplified in practical weather or material tables (psychrometric charts, Ashby material selection charts...) which need not be shown but are indeed helpful. Other conceptual documents, like Behling's triangles, accurately represent the need to prioritize issues related to form and matter that are essential to achieve a good thermodynamic performance from the beginning, which is important to later avoid prostheses that almost always evidence shortages in the original design, spoil the whole and make maintenance and construction more expensive.

The protocols we propose our students (see the text "Prototypes and Protocols") explain well some of the procedures that we have gradually individualized with the help of international experts. Once used it is hard to do without them. These procedures lead to what we have called "thermodynamic monsters" because of their additive character (they identify everything that is potentially operative in the project). Thermodynamics and architecture are synthetic and, just like engineering, they always prioritize the essential, in accordance with a dual process we refer to as "listing and ranking."

Analyzing the positive factors of the climate in each place and closely studying the most characteristic aspects of such climate (sometimes apparently negligible, like the night summer breezes in Madrid, the influence of which could go unnoticed on the charts, but whose efficiency "thermally restarting" the city every summer night is essential) is a ritual we never delay; it precedes every sketch or drawing we discuss, and is a necessary piece of knowledge to reflect upon organization, form or matter.

These initial aspects can occasionally give rise to projects of unforeseen reach. For instance, after analyzing Madrid's climate numerous times, a new study which paid more attention to its global significance unveiled that it is the only European metropolis that can rethink its housing and office market as one hundred per cent passive, something which in itself harbors a political proposal – to transform and refurbish the historic center, for instance – which cannot derive from quantitative studies. With this idea we developed a course at Harvard's GSD with Matthias Schuler, whose experimental content revealed the high potential of this hypothesis to reassess the conventional intervention techniques and methods, with effects on health and comfort (radical reduction of the heat island effect and environmental pollution) that point at other ways of addressing "sustainability," as it was until just recently, and that today opens up to create relationships between our disciplinarian culture and contemporary science. These objectives are obviously completely different from those of other more technocratic approaches.

When taking a look at what building energy certifications seek, we find contradictions that encourage us to avoid falling prey to their mercantile simplifications. It is impossible, for instance, to obtain a LEED certification using passive construction methods, because since there is no power use (air conditioning), energy efficiency cannot be "measured." This contradiction is even greater if we think about the role given to public space in those rating systems. Basically, they deal with internal efficiency, but they never say where the thermal gains go when released: to the public space. The more efficient the building's enclosure, the greater the chances that the street will become a thermal waste site, as happens with light, rear ventilated facades. Only by studying the impact of a building's thermal mass (and its location inside and/or outside the facade) on the comfort conditions of public spaces, can we think about a system to level both the public and the private sphere: the frontier between them is by no means neutral, and in fact it also acts in one or another direction. Another example: the impossibility of evaluating the beneficial effect of trees in cities (measuring the effect of tree shade, the cheapest and most efficient system, not to mention aesthetics or biodiversity, for comfort in the Mediterranean and tropical city) excludes them from any system based on quantifiable data. Though there is a growing tendency to include these aspects, their contribution to the design process is virtually zero right now.

**Creative Thermodynamics**

Without doubt, under challenging climate conditions (when there are humidity and high temperatures, for instance) or when project conditions make it hard to design the building adequately, or simply when we can try out new ideas with greater freedom, the help of professionals with scientific background is essential, and should be sought at the project's outset, particularly in the case of collaborators specialized in materials science and building physics. But aside from this it is necessary to develop a holistic way of analyzing the role of thermodynamics that should question in a global way the modern assumptions regarding what kind of building systems we should use, starting out with the air conditioning systems invented by Carrier and based on the transportation of cool and heat using air treated as basic fluid, a highly inefficient and unhealthy system that has led to a minimization of the volumes of air in buildings to economize cost and reduce the role of architecture to the mere manufacturing of insulating and sealed enclosures.

Architecture transformed into a refrigerator, as Kiel Moe accurately put it, is a brutal limitation with an evident capacity to eliminate cultural, environmental and social interaction aspects, because it stems from a principle that is radically opposed to the idea of city:

neutralizing as much as possible all exchanges with the exterior, whichever it may be.

Thermal mass has been replaced by thermal insulation; natural or convective ventilation, by the building's sealing. The only parameter of materials that are activated by this conception is conduction; factors like emissivity and diffusivity, important for thermal comfort, go unnoticed, because the inherited thermal analysis regime is static, the time and variations in the distribution of flows are absent, but the daily cycle is always changing, and the change in hygrothermal conditions is essential to comfort. Materials do nothing but decorate, but the reality is different. It is not a matter of simply questioning the importance of some of these inventions and systems, but of understanding that we possess the technical and cultural conditions needed to overcome them and create prototypes and building complexes in which both form and matter become that thermal machine; nothing new if we consider almost one hundred per cent of the vernacular architectures in the planet. However, these are no longer the example to follow, because the conditions we face aren't the same: neither the scale, nor the programs, nor the densities, nor the relationship with infrastructures, nor the globalized contemporary material culture allow looking back. Contemporary architects are orphans in two senses: orphans of history and almost orphans of modern forbears, subjected to regulations based on analysis systems that falsify reality. Resources must be understood and used in a different way, and this is possible now because there is enough data to get around what we have and don't like.

The mechanization of interior spaces (domestic or offices) brought about by the widespread use of digital tools has prompted to reevaluate the concept of interior thermal stress. A playroom for children, full of digital gadgets, can be compared to a tiny data center or to high-tech offices. The exponential multiplication of interior thermal loads due to the combined action of people and machines – in all the different areas of our lives – urges to manage and adjust them with the help of energy exchange rings. These, at the same time, must be related to other rings of exchange with the exterior, on a daily and seasonal basis. Mixed-use programs are not only a good commercial option in high-density cities, but also a formula still in a purely experimental phase to create open systems and dissipative structures that were until recently more appropriate to biology or ecology than to architecture.

### Architecture as Mediation

We could go on listing other matters that a qualitative conception of aspects related to the thermodynamics of buildings and cities brings up when considered in qualitative terms (that is, creative) and not exclusively quantitative terms. If we go back to the main statement of this text, we can understand that this knowledge is a critical, cultural and political weapon, but also creative in aesthetic terms and, that this is the most interesting angle from which thermodynamic principles can be considered in architecture. Holistic knowledge is close to the theoretical perspective, to the need to set a distance to see the whole picture. So, to sum up, we are particularly interested in the beginnings of the processes, in ensuring that the first decisions – those regarding form, mass, air, relationship with natural elements, management of internal and external thermal gains – are economically self-organized and function as starting point of the project, and are present before determining any given scheme. We are particularly interested in questioning assumptions about matter and construction inherited from a method filled with fallacies that should be in the public domain, because they involve specific policies that affect all areas of daily life. And we are particularly interested in the broad creative panorama that opens up when rethinking cities from a thermodynamic perspective, both on practical, theoretical or speculative level: how interpreting carefully one single deceptively trivial factor, it is possible to rethink the scales, methods, and objectives that we considered to be true. All this is applicable (with different degrees of perfection) to the academic field and to professional practice. In fact, they constitute a feedback system that is crucial in redefining the role that academics plays in our field, and in architecture's role mediating between scientific, cultural and social facets. Just like some landscaping professors forbid their students from using green as a preemptive measure (avoiding clichés), in our office or our classes we rarely draw the arrows that some expect to see, unless we are describing complex phenomena in which our capacity to visualize invisible dynamic phenomena is limited. Depending on how they are viewed, thermodynamic principles are concealed or apparent in the conglomerates of form and matter that interact with the natural elements that define the dynamic regime of our climatic environments. These material organizations take on most of the thermodynamic behavior in the absence of a finetuning that can only happen after the pertinent quantitative analyses have been done.

But these material organizations are in themselves Architecture, the purpose of our work, and are explained with simple formulas and principles, particularly when included in academic programs to advance in the construction of an architecture capable of expanding the limits of our experience and provide an answer to the challenges of 21st-century cities.

(To avoid misunderstandings: none of this questions the value of quantitative knowledge, and in fact it generates demands that are much more important than thinking whether two more centimeters of insulation will mean going from bronze to silver in the certification ratings, something that, though caricatured, is the focus of all environmental computer programs applied to architecture, as well as to their close relatives: energy certification stamps).

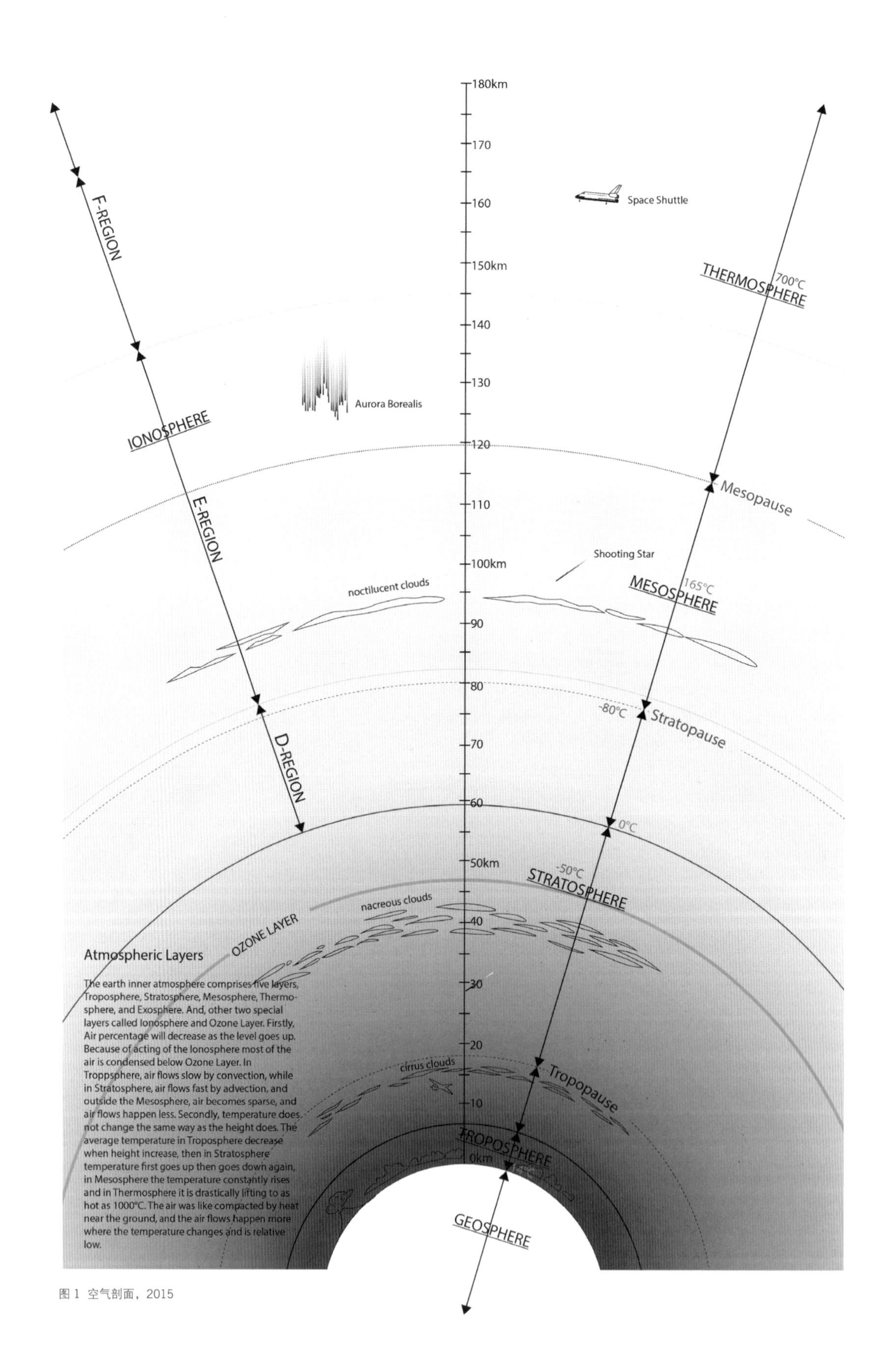
180km
170
160
150km
140
130
120
110
100km
90
80
70
60
50km
40
30
20
10
0km
Space Shuttle
Aurora Borealis
Shooting Star
F-REGION
IONOSPHERE
E-REGION
D-REGION
THERMOSPHERE
700°C
Mesopause
MESOSPHERE
165°C
noctilucent clouds
-80°C
Stratopause
0°C
-50°C
STRATOSPHERE
nacreous clouds
OZONE LAYER
cirrus clouds
Tropopause
TROPOSPHERE
GEOSPHERE
Atmospheric Layers
The earth inner atmosphere comprises five layers, Troposphere, Stratosphere, Mesosphere, Thermosphere, and Exosphere. And, other two special layers called Ionosphere and Ozone Layer. Firstly, Air percentage will decrease as the level goes up. Because of acting of the Ionosphere most of the air is condensed below Ozone Layer. In Troppsphere, air flows slow by convection, while in Stratosphere, air flows fast by advection, and outside the Mesosphere, air becomes sparse, and air flows happen less. Secondly, temperature does not change the same way as the height does. The average temperature in Troposphere decrease when height increase, then in Stratosphere temperature first goes up then goes down again, in Mesosphere the temperature constantly rises and in Thermosphere it is drastically lifting to as hot as 1000°C. The air was like compacted by heat near the ground, and the air flows happen more where the temperature changes and is relative low.

图 1 空气剖面，2015

# 经由研究的设计

周渐佳

（建筑是）无人之岛，且它的边界永远在变化。

——曼弗雷多·塔夫里，1979

## 导言

“热力学建筑原型”是同济大学建筑系开设的面向本科四年级学生的自选题课程设计。在此之前，本科学生接受的设计课程训练大多围绕着最基本的功能、场地、构造等问题展开，也更趋于关注建筑本体。与这些课程相比，“热力学建筑原型”无疑更具实验意味，在学生具备基本的设计能力的基础上，强调方法论在推进设计过程中起到的作用。究其原因，在我们面对“热力学建筑”这样一个集合了诸多其他学科知识与研究方法的议题时，传统的设计理论已经无法以单一轨迹的方式发生作用，这使得我们需要借助的工具、需要掌握的知识更加多样化；另一方面这些来自其他学科的知识对建筑本体所产生的影响并非只流于表面。这也意味着，当建筑面临的问题趋于复杂时，建筑生产的边界在不断扩大，而建筑本体的问题可能会发生根本性的转变。“热力学建筑原型”正是试图从能量的维度出发，重新看待建筑生产的方式与结果。

在特定的历史时期中，建筑师与理论家、历史学家们尝试通过建立某种共识性的话语去理解社会、科学和经济的发展给建筑学科带来的变革。20 世纪六七十年代的动荡加上新兴历史意识的觉醒，使得各种各样的建筑与城市理论随之涌现，这些理论都试图提供一个包罗万象的解答。然而也只有经历过这样一个阶段，才可能看到其问题所在——这种试图以“理论”解释一切的立场其实有很强的约束性，一方面其对城市的经验是分散的，另一方面理论问题中也存在着某些相对性。进入 20 世纪 80 年代之后，哲学思想的引入更是极大地扩展了建筑话语本身的丰富性，也是在同一时间段内，我们对建筑的知识与工具有了更明晰的认识，建筑学科的核心，或者说“自治性”被进一步确立。而跨学科与多学科不断介入建筑学科的外缘的过程，也在不断澄清建筑学科在更广泛的实践话语中的地位，同时开启了对学科之间共享领域的讨论。从这些探索中走出来的是，“研究”作为一种独特的建筑实践方法，独立于设计、理论、历史和批评以外而出现。

20 世纪 60 年代曾经出现的设计方法运动，无论结果如何，希望在设计与研究之间建立一种新的关系，“设计方法”被视为一种为了得到设计解决方案而建立的系统化的过程。但是回溯其发生的社会背景，对科技进步的乐观主义无疑推动了研究与大规模生产之间的紧密关联，以及试图让设计过程更加“科学”的驱动，因此对设计方法的普遍讨论往往出现在科技与制造业极速发展的时代。反观国内近几年的趋势，也许可以被视为“热力学建筑原型”以及整个设计—研究讨论出现的注脚。

## 理论 vs 研究

事实上，“研究”或者“设计—研究”是不同于建筑理论的实践范式。从《建筑十书》开始，“建筑理论”以论文写作为主的伟大传统就已确认，它与“制图”一起，形成了建筑的最重要的两种实践范式。“建筑理论”将建筑实践的定义进一步扩展如下：对知识的收集与排序，以及对实践中产生的内在逻辑与意义的辩论与讨论。直到 18 世纪，建筑理论的任务都是维持并巩固这种既定的特点——以一套系统化知识，加上一套指导法则为建筑生产确立基础。理论被视为在建筑中持续思考、工作的有效方法。而随着时间的推移，理论开始越来越强调自身反思的角色，逐渐演变出建筑批评的分支，也更趋向于成为对当代情境的反应。

然而，当代理论的局限也逐渐显现。事实上，是大量过往案例决定了建筑理论自身的倾向，这也决定了理论中一大部分工作仍然是处理以往的经验，当理论所能覆盖的案例的数量、种类都开始变得有限时，就意味着理论已经不能如预期那样全盘指导建筑的行动。尤其在建筑行业高度分工的当下，理论已经与其所应对的境况有所疏离，也无法提供系统的知识体系。同时，建筑师所承担的工作也一再经历泛化与扩充，变得更加开放与灵活。换言之，理论的工作不再是指导，而是尽可能地趋近建筑生产的现状本身。因此，近几十年，研究成为理论之外的另一种方法，提供了一套更加具体的工具，也与多样的建筑生产更加契合。当以往的建筑理论难以在当代建筑中发挥作用时，研究扮演的角色使基于写作、调研的工作与设计更为密切地结合在一起。无论在教学还是在实践过程中，研究希望确立的是特定类型的知识，以此作为设计的背景，这其中包括与特定类型或特定地点相关的社会、政治、经济状况，在“热力学建筑原型”中特别指项目所在地的气候与能量状况，这些通过研究获得的信息决定了特定的前提条件。这些前提条件的作用接近于勾勒出整个话题的知识背景，并且以此为线索找到合适的案例、地域以供借鉴。真正的困难在于，无论是通过理论还是通过研究所得到的知识，如何以工具化、可操作的方式来直接或间接地影响建筑设计的过程？毕竟我们最终希望引导的是经由研究而得出的设计。

## 再现

在课程伊始，制图是带领学生进入研究语境的重要手段之一。作为建筑学科自主性的标志，我们认为有必要重新发现制图，或者重新发现呈现的方式，鼓励学生通过基础、传统甚至直白的建筑图纸来完成对研究内容的再现训练。对于“热力学建筑原型”而言，四个学期的课程设计分别设置了空气、光、热、能量等主题，这些不可见、动态的对象需要学生通过再现去捕捉。事实上，制图的过程是对知识进行重新排序与组织的过程，这个目的与研究的初衷相仿，可以类比于训练学生通过特定的透镜去审视城市、理解基地、阅读案例，并最终找到适当的方法进行再现，而不是无意义地生产图纸。在设计推进的不同阶段，制图会呈现出不同方向的重要性，有时甚至有必要为之重新创造一种建筑的视觉语言。

例如在基础性的研究阶段，课程会要求学生通过全球地图或者大剖面来建立对讨论对象的了解，其中包括 2015 年绘制的从地表到大气层的空气剖面，距离、密度、温度等信息都在图中得到呈现（图 1）；从 2016 年开始，几乎每年的课程训练中都会通过合作的方式完成一张与主题相关的世界地图，反映不同温度带、植物带在全球范围内的分布（图 2）等。进入案例阶段后，学生们通过自己建立的线索来挑选一个或多个参考案例，并且对案例中的建筑形式进行有意识的抽象、归纳与重新表达，我们始终鼓励学生带着一种整体性的视野去理解案例，从中得出不同于以往的结论。这其中既有地域性非常强的建筑，例如希巴姆古城、印度建筑师多西的作品，也有更具一般属性或依赖技术手段调控环境的建筑，例如柏林自由大学图书馆。从这个阶段开始，模型被逐渐成为再现的手段之一，并为接下来的原型阶段做准备，由此逐步建立建筑形式与物质、能量的关联。事实上，再现不仅是对研究内容的演示，更是从热力学维度重塑建筑动机的一种方式。

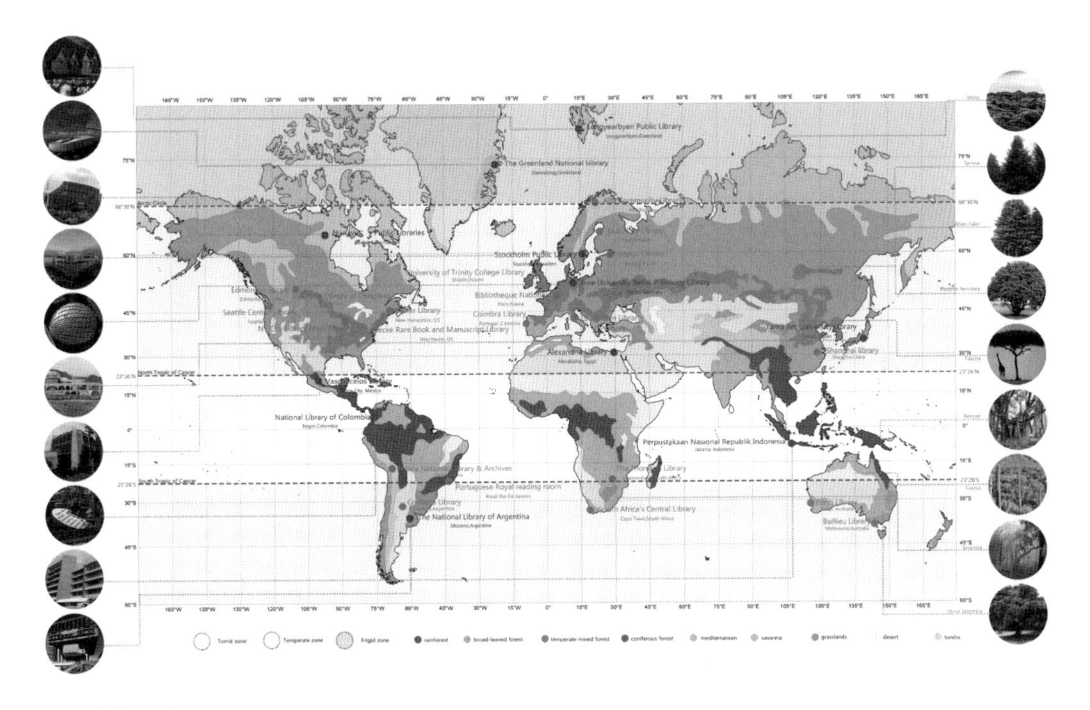


图 2 全球植被分区地图，2016

## 研究 vs 设计

引用马克·舒德比克在《地图与建筑设计中的实验》一文中的观点，建筑设计与建筑研究之间的关系可以被归纳为如下三种：一是把建筑视为研究的特殊形式，因此设计本身就是一种调研行为；二是把设计视为研究的对象，通过把设计过程视为方法论的施用，从而把设计描述为一系列合理控制的步骤；三是厘清研究是如何潜在地影响设计，从而让空间维度的调研与设计行为直接发生关联。在“热力学建筑原型”的训练中，我们更倾向于用第三种分类来理解研究与设计之间的关系。那么在这个过程中，研究扮演的究竟是怎样的角色呢？通过课堂上不断的讨论与表达，我们发现研究实质上是创造一种集体的知识，这种集体的知识是可以被交流与分享的。而这种集体性是在对抗设计中一直以来的直觉、灵感的那个部分，是在对抗将设计与个人创造划上等号的行为，也是努力将设计的结果与更广义的社会背景相关联的隐喻。从这个意义上来说，研究或者经由研究的设计既是建筑知识生产的方式，也是实践之外的建筑生产的重要方式。

在研究的语境下，建筑的问题与越来越多的学科产生交集，项目所在的地域与文脉在很大程度上影响了建筑生产的背景，而潜在的哲学、文化、政治和美学价值决定了建筑生产的意义，但是这些以文本、地图、图标等为主的研究成果与建筑设计仍然存在着学科与工具上的差异。建筑在科学、文化和社会之间扮演了居间的角色，也为重新看待城市与地域环境提供了新的视角，这并不局限于实践或理论，完全可以是实验性的。当我们回顾这些知识的时候，可以将其视作批判的、文化的和政治的武器，但是它们同样有着美学意义上的可能性，这也是设计—研究上特殊且有趣的一点。伊纳吉·阿巴罗斯在论述中提出，形式在建筑与外界环境的能量交换过程中扮演了重要的角色，并且进一步用三角形图解去诠释热力学维度下环境、技术与形式之间的关系，在他看来，形式也就是三角形的基座应当成为气候适应的基本原则，也是从研究到设计的转译过程中最重要的步骤。

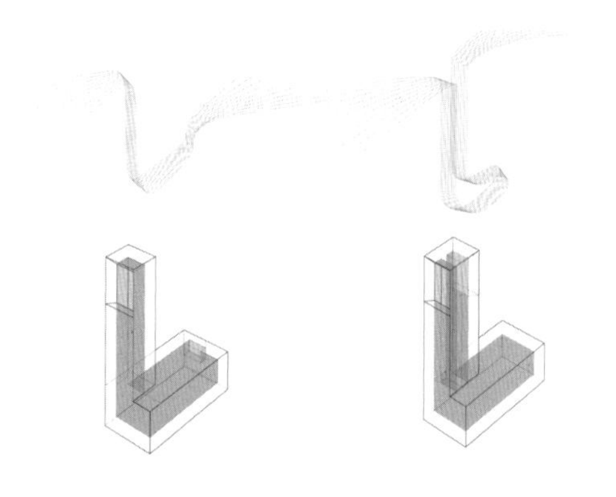

图 3　风塔原型研究，张润泽，2015

## 原型

热力学的原则在形式与物质之间或显现或隐藏。这里的形式是辩证的，它既包含了经过历史经验证实并且获得某种共识的典型，也包含了带有科学性、实验性的原型。如果说前者是研究阶段的对象，那么后者则是经由研究的设计所得到的成果。原型的介入很大程度上改变了设计过程中需要理解的设计目标，它带有抽象的属性，而这种抽象性的本质是学生设定的一种论点或一种假设，是将复杂系统还原到最初始的、最根本的状态，通过不断的检验、优化以确立原型本身的有效性。原型之所以重要，是因为它一方面构成了从研究到设计最重要的一步跨越，以建筑的语言去表述知识；另一方面原型自身是可类比、可操作的。而在阿巴罗斯看来，原型同样有美学上的意义，提供了产生全新形式语言的可能。原型同样可以是一种参照，以此重新评估建筑传统中的技术与方法，甚至反思各个构件的效用，这一点在热力学建筑的背景下显得尤为重要。课程中涉及的案例构成了一种典型，但是学生们必须站在特定的立场去重新拆解这样的典型，将其中最重要的原理抽取出来，再发展为自己的原型。

原型的生成同样不受限于尺度或者范围，在教学实验中，既出现过以中东地区建筑中作为附属结构的风塔作为研究对象（图 3），也出现过以自然村落的集群作为研究对象（图 4）。一定程度上，这也体现出原型在尺度上的自由性，这种自由性也正是来源于气候、能量这样的强度特质。学生们可以用不同的模型、制图、模拟去测试原型的可能性，由此带来的原型设计也是多样的，有些以生成的形态作为原型，有些着重讨论体量之间的关系，有些则是通过实体去塑造物质流通的路径。但最重要的是，原型是一种可以理解、可以讨论的对象，也是把研究转化成设计的演练过程，并最终落回到建筑学科的语境中来。随着原型被植入基地和地域环境而变得越来越具象化，它将实现自身从抽象模型到建筑的转变。

## 结论：一种范式的迁移

设计—研究在欧洲和美国已经有其传统，虽然时间不长，却是在社会历经转变时的必然产物；而在中国它没有任何先例可言，完全是来自其他国家与院校的植入，这种状况也决定了以研究、原型为导向的教学实验尚在少数，且无定式可言。不过这也造就了这样一个话题本身的开放性。不同阶段的评图，会邀请建筑师、理论家、工程师、科学家等不同学科背景的嘉宾参与，由此形成的开放式的反馈也是教学实验对热力学原型建筑、对当代建筑境况的反馈，这种高度的开放与交叉也正是研究本身所追求的。建筑学中试图指导全盘行动的理论已然失效了，必须通过更有针对性也更为综合的知识体系去回应复杂的问题。

这带来的必然是建筑工具与方法论的多样化，并且必须通过再现、原型之类的媒介去完成不同工作方法之间的转译，最终形成建筑化的语言。在这个过程中，研究所涉及的包括社会、地域、文化、气候在内的外界环境与建筑的本体同样重要，且密不可分。这也意味着仅关注建筑本体状态的终结，或者说设计对象从建筑本体转变为与之关联的环境。那么研究是否会成为实践与理论之外又一种建筑生产的范式？通过诸如“热力学建筑原型”这样的课程，我们是否可以甄别出设计—研究中有效的方法论？如果是，那么这些方法论的目的、性质、有效性以及在定性、定量意义上的成果又是什么？更重要的是，研究对于新的建筑知识的生产，对于设计思维的方法是否会有新的贡献？也许建筑和建筑的知识已经悄然进入新的发展阶段。

参考文献

[1] Manfredo Tafuri. Main Lines of the Great Theoretical Debate over Architecture and Urban Planning, 1960-1977[J]. A+U, 1979（01）:142-161.

[2] Iñaki Ábalos. Why we don't draw arrows (hardly ever)[J]. AV Monographs 169, 2014: 26-31.

[3] Johan De Walsche, ed. Prototype and Diagram[M]. TU Delft Open, 2016.

[4] Jeremy Till. Three Myths and One Model[EB/OL]. (2017-01-03) [2018-09-30]. https://jeremytill.s3.amazonaws.com/uploads/post/attachment/34/2007_Three_Myths_and_One_Model.pdf.

[5] 李麟学 . 热力学建筑原型：环境调控的形式法 [J]. 时代建筑 , 2018（03）: 36-40.

[6] 鲁安东 . “设计研究”在建筑教育中的兴起及其当代因应 [J]. 时代建筑 , 2017（03）: 46-49.

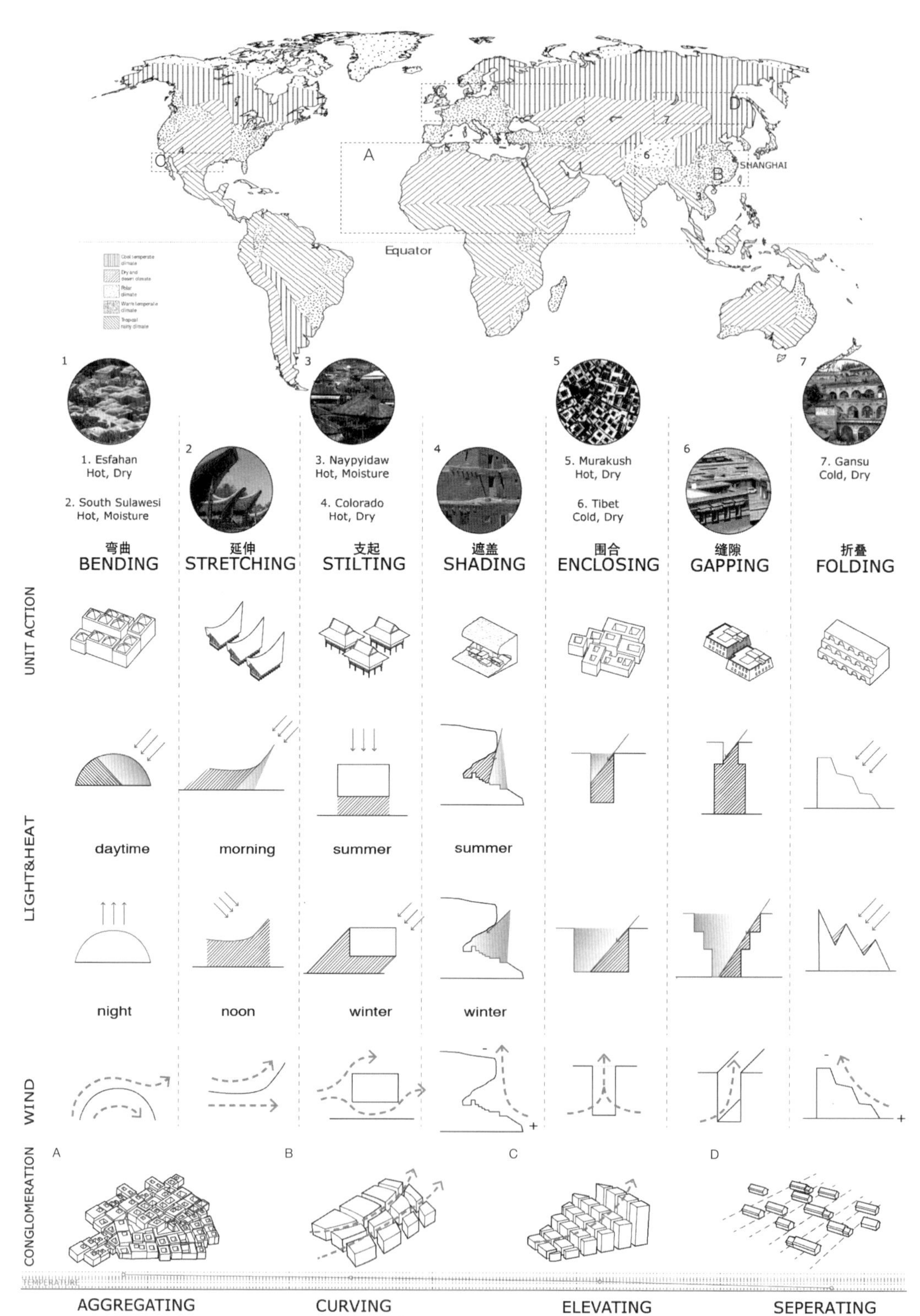


图 4 自然聚落研究，梁芊荟、林静之、王劲凯，2016

# Design by Research

**ZHOU Jianjia**

Architecture…is a "no man's land," the boundaries of which are forever shifting.

— Manfredo Tafuri, 1979

### Introduction

"Thermodynamic Architecture Prototype" is an elective design studio for the 4th year undergraduates offered by the Department of Architecture, Tongji University. Prior to this, the design course training for undergraduate students mostly focused on the basic issues such as program, site, structures, etc.. Compared with these courses, "Thermodynamic Architecture Prototype" is undoubtedly more experimental. On the basis of student's design ability, it emphasizes the role of methodology in developing design process. When we are faced with "thermodynamics," the topic that integrates the knowledge and methods from many other disciplines, conventional design approach can no longer function in a single trajectory; instead, more diverse tools and knowledge are required. On the other hand, the influence of this knowledge from other disciplines on architecture ontology is substantial. When problems faced by architecture discipline tend to be complicated, the boundaries of architecture production are expanding, leading architecture ontology to fundamental changes. "Thermodynamic Architecture Prototype" attempts to revisit the way and result of architecture production from the lens of energy.

In specific historical stages, architects, theorists, and historians tended to understand the changes brought to architecture by social, scientific and economic forces by establishing some common discourse. The upheavals of the 1960s and 1970s, coupled with an emerging historical awareness, led to the flourish of architectural and urban theories, all attempted to provide an all-encompassing theory of architecture. However, it is only after experiencing such a stage that it is possible to see the problem - this position of trying to explain everything by "theory" is actually very limited in scope, partially because of the fragmentation of the experience of the city, and partially due to the relativism in theoretical problems. After the 1980s, the introduction of philosophy greatly expanded the richness of architectural discourse, meanwhile, a clearer understanding of knowledge and tools of architecture was built up, in other words, the core of this discipline, or "autonomy" is further established. The process of inter- and multi- disciplinary investigations allowed for an assessment of the outer edges of the architecture discipline, which not only clarified the position of architecture in the larger filed of discursive practices, but also opened up the discourse to the shared fields. "Research" emerges as a unique architectural practice from these explorations, independent from design, theory, history, and criticism.

In the 1960s, there was a design method movement. Regardless of the outcome, the movement wanted to establish a new relationship between design and research. The "design method" was considered as a way to get a design solution via a systematic process. Indeed, the notion of "design methodology" typically came into existence in the era of technological advancement and manufacturing from the beginning of the 20th century onwards. Under the background of rapid development of technology and manufacturing in China, this course can be regarded as a footnote for this tendency.

**Theory vs. Research**

In fact, "research" or "design-research" is a practical paradigm different from theory. Starting from Vitruvius's *Ten Books on Architecture*, the origin of "architecture theory" has been located in the tradition of treatise writing. Together with "drawing," these two form the most important paradigm of architecture production. The architectural theory further extends the definition of architectural practice to the following two: the collection and sequencing of knowledge, and the debate and discussion of the inherent logic and meaning generated in practice. Until the 18th century, the task of the architectural theory was to maintain and consolidate this established feature —establishing a foundation for building production with a systematic knowledge and a set of guiding principles. The theory is seen as an effective way to continue thinking and working in architecture. With the passage of time, the theory began to emphasize its reflective role, gradually evolved into a branch of architectural criticism, and more responsive to contemporary situations.

However, the limitations of contemporary theory are gradually emerging. In fact, a large number of past cases determine the tendency of architectural theory itself, which also determines that a large part of the work in the theory is still dealing with past experience. When the number and types of cases that the theory can cover begin to become limited, it means that the theory has not been able to guide the architectural actions as expected. Especially in the current high division of labor, theory is ill-equipped to provide a systematic knowledge system. At the same time, works undertaken by the architects become more open and flexible. In other words, the theoretical work is no longer a guide, but rather as close as possible to the status quo of architecture production itself. Therefore, in recent decades, "research" has become another method beyond "theory," providing a more specific set of tools and methods, and more in line with diversity. While previous architectural theories were difficult to play a role in contemporary architecture, research has become a more closely integrated work based on writing and research with design. Whether in teaching or in practice, research seeks to establish a specific type of knowledge as a context for design, including social, political, and economic conditions associated with a particular typology or sites, which in "thermodynamic prototypes," refers to the climate and energy status of the project site, and the information obtained through the research determines the specific preconditions. These preconditions outline the knowledge background of the whole topic, and are used as a clue to find suitable cases and regions for reference. The real difficulty lies in whether the knowledge obtained through theory or research can directly or indirectly affect the process of architectural design in an operational way. After all, what we ultimately hope to achieve is design derived from research.

**Representation**

At the beginning of studio, drawing is one of the important means of leading students into the context of research. Being the sign of the architecture autonomy, we believe it is necessary to rediscover the drawing, even the way of presentation, and encourage students to explore through basic, traditional and even straightforward architectural drawings. For this course, themes such as air, light, heat, and energy are set in the four-semester period.

These invisible and dynamic objects await to be captured and visualized. In fact, drawing is a process of reordering and organizing knowledge, similar to the purpose of research. It can be analogized to training students to examine the city through a specific lens, then understand the site, read the case, and finally find the appropriate method instead of producing meaningless images, sometimes it is necessary to invent a visual language.

For example, in the research phase, students are asked to establish an understanding of the object of discussion through a global map or a large section, including an "air section" from the ground surface to the atmosphere drawn in 2015, showing the information of distance, density, temperature, etc. (Fig. 1); From 2016 on, students need to produce a world map related to the assigned theme through cooperation, reflecting the distribution of different temperature zones and plant bands on a global scale (Fig. 2). In the case phase, students select one or more reference cases following their own clues, to abstract, summarize and finally represent the architectural forms in this case. We always encourage students to take a holistic approach to understand the case, from which to draw different conclusions from the conventional ones. Among them are some significant architecture and architects, such as the ancient city of Shibam and the works of the Indian architect Doshi, or even more generic buildings that rely on technical means to control the environment, such as the Free University Library in Berlin. From this stage, model is gradually introduced into design development, also prepares students for the next prototype stage, thereby establishing the connection between the architectural form, matter and energy. In fact, representation is not only a way of demonstrating research content, but also a way to reshape architectural motives from the thermodynamic dimension.

**Research vs. Design**

Citing Marc Schoonderbeek's view in "Mapping and Experiment in Architectural Design," the relationship between architectural design and architectural research can be summarized as follows: (1) to consider design as a specific form of research, thus considering the act of design in itself as an investigative act; (2) to consider design as object of research, by concentrating on design as methodological process, thus describing design as a reasonably controlled procedural act; or (3) to clarify how research might potentially inform design, thus directly relating spatial investigation to the project act of design. It should be clear that "Thermodynamic Architecture Prototype" focuses primarily on the third category. So what role does "research" play in this process? Through continuous discussions, we find that "research" is essentially creating a collective knowledge that can be exchanged and shared. And this "collectiveness" is crucial to guarantee that architecture is more than a matter of intuition or the act of equating design to the personal signature. It is also trying to relate the result of design to the broader social background metaphorically. In this sense, "research" or design through research is both a means of producing knowledge and of architecture production outside of practice.

In the context of research, the problems of architecture are intertwined with more disciplines. The geographical and cultural contexts constitute the background of architectural production, while the underlying philosophy, culture, politics, and aesthetic value determine the meaning of it. But these research outcomes based on texts, maps, icons, etc. remain alienated in terms of architectural language or tools. Architecture plays an intermediary role between science, culture, and society. It also provides a new perspective for revisiting the city and local environment. This is not limited to practice or theory, and can be speculative. Design-research has a distinctive feature that not only can it be seen as a critical, cultural and political weapon, it also has the aesthetics potential. Iñaki Ábalos affirms that form plays an important role in the energy exchange between architecture and the external environment, and he uses triangle diagrams to interpret the relationship between environment, technology, and form in the thermodynamic dimension. In his view, the base of the triangle - the form - should be the basic principle of climate adaptation and the most important step in translating from research to design.

### Prototype

The principle of thermodynamics is either apparent or hidden between form and matter. The form here is dialectical. It contains both "typical," meaning historically proven ones, as well as scientific and experimental ones as "prototype." If the former is the object of the research phase, then the latter is the outcome of the design by research. The involvement of "prototype" largely changes the design goals that need to be understood in the design process. It has abstract attributes, and the essence of this abstraction is an argument or a hypothesis set by students. It is to restore the complex system to the most fundamental state, through continuous testing and optimization to establish the validity of the prototype itself. The prototype is important because it constitutes the most important step from research to design. It expresses knowledge in the language of architecture. On the other hand, the prototype itself is analogous and operable, but in Ábalos's argument, the prototype also endows the possibility of producing a new formal language. Prototypes can also be a reference to re-evaluate the techniques and methods of architectural conventions, and even reflect on components, which is particularly crucial in the context of "thermodynamic architecture." The cases involved in the course constitute a "typical," but students must stand in a specific position to deconstruct such "typical," extract the most important principles, and develop their own "prototypes."

The generation of prototypes is also not limited to scales or scopes. In the teaching experiments, cases like wind towers with auxiliary structures in buildings in the Middle East (Fig. 3), or clusters of natural villages were all covered (Fig. 4). To a certain extent, this also reflects the freedom of the prototype in scale. This freedom is also derived from the intensive nature of climate and energy. Students use different models, drawings, and simulations to test the possibility of prototypes, resulting in diverse prototypes. Some use the generated form as a prototype, some focus on the relationship between masses, and some on shaping the path for the matter to circulate. But most importantly, the prototype is an object that can be understood and discussed, as well as a trigger that transforms research into design, and eventually falls back into the context of architectural discipline. As prototypes are embedded in the site and territorial environments, as they become more visible, prototypes that implanted are capable of transforming themselves from abstract models to architecture.

### Conclusion: A Paradigm Shift

"Design-research" has a history in Europe and the United States, though not long, but it is an inevitable outcome of the transformation of society. Here however it has no precedent , as something entirely implanted from other countries and institutions. That explain why determines that research and prototype-oriented teaching experiments are still in the minority, and there is no fixed formula. But this scarcity also mean an openness on such a topic.

In the review stages, we invited architects, theorists, engineers, scientists and others from different disciplines to contribute, thus forming open feedback. It is also the feedback of teaching experiments on "thermodynamic architecture prototype" and the situation of contemporary architecture. This high degree of openness and intersection is exactly what "research" itself pursues. All-encompassing theories in architecture have lapsed and must respond to complex problems through a more targeted and integrated body of knowledge.

This will inevitably lead to the diversification of building tools and methodologies, and through the mediation of representation and prototypes to complete the translation between different working methods, and finally form the language of architecture. In this process, the external environment involved in the research, including society, geography, culture, and climate, is as important as the ontology of the architecture. This also means a shift of focus, from the state of architecture as an object to the environment associated with it. Will "research" become a paradigm of architecture production in addition to practice and theory? Can we identify effective methodologies in design-research through experiments such as "Thermodynamic Architecture Prototype"? If so, what are the purposes of nature, effectiveness in the qualitative and quantitative sense of these methodologies? More importantly, does research have a new contribution to the production of new architectural knowledge and to the way of design thinking? Perhaps the knowledge of architecture and the discipline itself have quietly entered a new stage of development.

# 基于知识的设计教学：一次关于环境、结构与地形的实验

谭峥

## 关于“物”与“流”的知识

建筑学是一门关于人类如何干预并重构物质世界的学科，环境、结构与地形是在建筑学语境中与物质世界直接联系的领域。“环境”关注建筑协调人与自然关系的中介功能；“结构”关注建筑平衡形式与内容关系的整合功能；“地形”关注建筑作为整个地表组成部分的连接作用。从广义上来说，这三大领域都能归结为“环境”，后者能够涵盖一切与管理并引导空间体验有关的知识。

除了关于政治与文化表达的知识，建筑学的知识基本上不会脱离上述三类领域。然而在当前的建筑学教育中，关于环境、结构与地形的知识分属于不同的课程门类（建筑结构、建筑技术、环境工程、场地设计、景观学，等等），并未有机地贯彻在单次建筑设计训练的整个过程中，学生也缺乏机会在一个设计中应用并检验关于物质运作规律的各类知识。基于此，急需构建一门着眼于“环境”的、以研究结合设计为要旨的实训课程，以弥补现有培养体系的不足。

张永和与谭峥所领导的教学小组从 2015 年开始“基础设施建筑学”的课程实验，将设计任务对象从传统认知中的“建筑物”转移到更广泛的城市基础设施（或“构筑物”）中，鼓励学生重新学习各种关于物质、能量与结构的知识，以应对真实世界中层出不穷的设计挑战。该项目以高年级本科生或研究生的研究型设计课程为依托，相继组织了“基础设施建筑学”（2015）、“滨水基础设施”（2017）与“青浦桥舍”（2018）等课题，并且进行了一系列的配套国际研讨会、专题讨论会和实地调查，以构想能源、机动性和人居条件的未来可能形式。

## 基础设施与环境

现代的基础设施网络已经在地球表面形成了一种新型人工地形，其意义应该从建筑学、人类学和地质学等角度重新审视。在基础设施高度发达的当下，人居环境越来越受制于各类技术条件。基础设施决定着人们的空间体验的度量——热舒适、满足感、机动性、可达性、安全性，等等。这种对“现代性”的体验越来越脱离文化与习俗的惯性，成为一种普适的要求。然而，解决住房危机、贫困和社会流动性问题的钥匙仍然在那些掌控资金、劳动力、能源和所有自然资源的各类“利益相关方”手中。所以，相对抽象的“基础设施”是由具体的系统、网络、机制、机构与构筑物构成的。建筑学介入基础设施的方法不应该仅仅是通过理解其技术运行规律，也应该是通过理解它的空间、组织及其背后各类机构力量的博弈。

基础设施是一种分配信息、能源、交通等公共资源的人工构筑物体系。在宏观尺度上，它可以表现为各种网络与系统；在微观尺度上，它表现为一种与日常生活密切相关的功能部件或可驻留的公共空间。当代的基础设施积极地介入现代生活形态，并将传统的街道、广场、市场纳入其体系，将自身转化为新的公共空间。以各种通道、枢纽、厢体为代表的交通基础设施空间已经

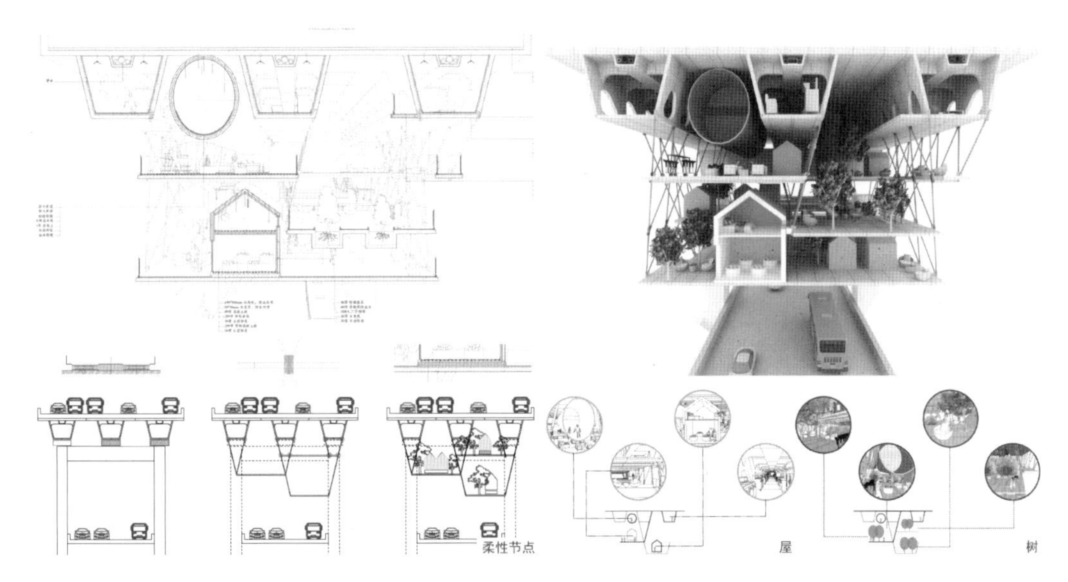


图 1 “基础设施建筑学”课程之“高架悬垂公园”方案，2015

成为具有集体记忆的场所，而信息与文化基础设施构成的虚拟公共空间也呼之欲出，基础设施的机械性面目已经开始柔化。它构成了本雅明所说的不可逃脱的“内部”[1]，并已经开始摆脱冷峻、隐匿的形象，成为社会生活中影响意义、符号与情绪发生的关键要素。相应地，关于“物”与“流”的知识体系也应当经过消化与转译，从传统的纯工程知识进入空间规划与设计所需要的基础修养基底。

## “桥舍”设计实验

“桥舍”毕业设计课程开始于2018年春季，是一系列以“基础设施建筑学”为主题的课程的延续，由同济大学与哥伦比亚大学联合开展。“基础设施建筑学”试图重置所有生产要素的消耗和增殖方式，探索在城市所能提供的基础设施条件之上设计更高标准的住居形式，以部分实现能源与劳动力的自主可持续利用。这些基础设施包括交通、能源、物流等系统。基于此，同济大学和哥伦比亚大学分别在上海和洛杉矶运作两个平行的设计教学计划，以测试：①流动人口住宅计划的策划；②创新性的城市住宅形式；③相应的基础设施解决方案。在具体的教学计划设计上，加州的教学计划以目前正在进行的城市救助站计划为背景，计划的核心是改造洛杉矶已有的废旧城市构筑物设施（仓库、车库等），使之成为被救助人员重归社会的家园，同时关注最终完成设施的能源自主、财务平衡、户型组合与模块化预制装配建筑技术的应用。上海的教学计划则考虑将低收入安置住所与新建的城市桥梁结合，同样应用多种节能与绿色建筑技术，并考虑最终建成环境与城市景观系统的融合。

“桥舍”基地位于上海市青浦区青浦大道，主体为跨越淀浦河的桥梁，桥梁连接淀山湖大道

---

1 本雅明在《拱廊计划》中将拱廊街视为一种室内外融为一体的内部（intérieur）。他认为拱廊是一种模糊且矛盾的空间复合体（将通道包裹在建筑物内），也展示了一种新型的住居方式。

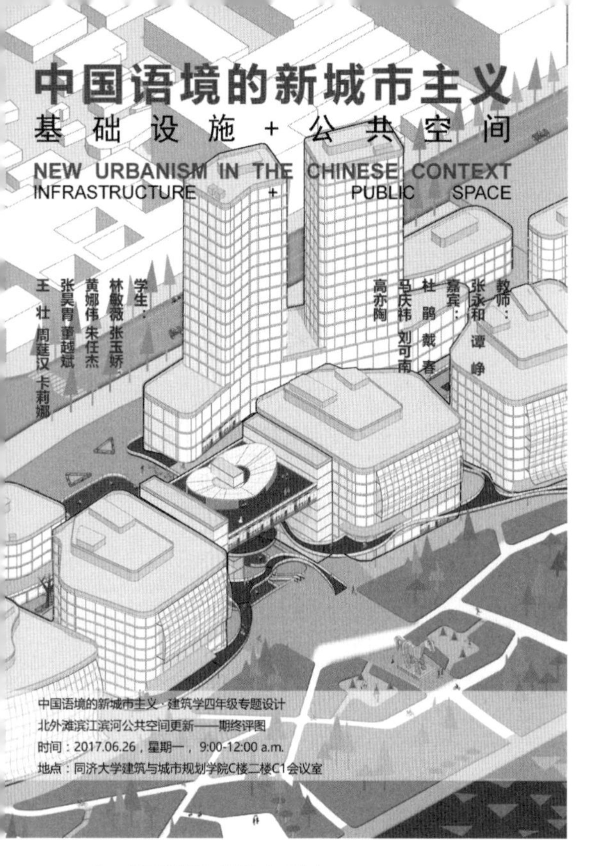


图 2 “基础设施建筑学”之“滨水基础设施空间”课程实验的评图海报，2017

与沪青平公路，淀山湖大道上有上海市轨道交通 17 号线上的“淀山湖大道”车站（17 号线目前已经开通）。本教学团队的张永和教授在基地上已经设计了一座步行与车行完全区隔的公路桥梁，该桥已经在施工过程中，受该桥梁形式启发，教学团队将这座桥的任务书演绎为可以安置低收入人群工作、住居的“桥舍”计划。桥舍首先需要满足且适当高于既定的桥梁标准，而且必须在河道蓝线范围内容纳一个流动人口安置站，以安置 100 个可供低收入人群工作及居住的预制装配式单元。桥舍的能源应当部分自主，并提供低收入人群的劳动空间，以实现社区能源供应的自足与人力价值的再生产。

本课题要求在项目中考虑：

1）该项目的主体为一座满足机动车、非机动车与行人通行的桥梁。由于河道为 VI 级通行航道，桥梁的通航孔净空尺度为：净宽不小于 30 米，上底宽不小于 22 米，净高不小于 4.5 米。

2）原规划青浦大道的道路红线宽度为 50 米，由于实现快慢分行，在不影响功能的前提下，实际桥梁宽度可以大幅缩窄。课题要求桥梁的机动车、非机动车与步行通行系统与周边环境良好接驳，不同的交通模式间有效区隔。“淀山湖大道”站周边区域在规划中为青浦区的新城市副中心。“桥舍”应当考虑慢性交通系统并考虑远期与“淀山湖大道”副中心区域接驳。可能的情况下可以在合理的想象下规划新型的交通工具与系统。

3）项目必须包含能够容纳 100 位低收入流动人口的临时安置站，也必须对这些居民进行生产组织，以实现生活自足。安置站的总建筑面积不超过 1500 平方米，流动人口的居住形式可以采用微型的预制装配式居住单元（即所谓的胶囊），必须包含必要的起居、卫浴、就餐、交往与管理设施。临时安置用房的出入口、交通、后勤应当与桥梁的日常通行互不干扰。

4）项目的能源消耗必须尽量自给自足，考

虑新型太阳能光伏技术与相关的节能技术来进行能源供应，考虑以雨水与中水设施来进行卫浴设备的清洁与景观植被系统的灌溉，等等，项目的景观系统必须与周边景观交融。考虑在一定的技术支持下（结构评估与优化软件）对桥梁的结构安全性进行基本的结构计算。

“桥舍”设计计划在运行过程中，以设计课带动相关设计知识的教学，将一般在研究生阶段才会出现的“读书会”形式纳入前期教学计划，要求同学阅读太阳能建筑设计、预制装配技术、道路交通设计与住宅设计的专业著作，随后撰写并讨论读书报告。与此同时，邀请环境、建造与结构领域的专业人员参与教学，以充实小组所需的设计知识与技能。课题小组分别在课程中期汇报与终期汇报以前，组织与纽约哥伦比亚大学的联合工作营以及与德州理工大学的非正式合作交流。

课题的结论与反思如下：

类似“桥舍”的空间类型在现代建筑史中不乏先例，在1950—1970年代流行的巨构建筑运动推崇在高速公路等基础设施上架设城市的设想，典型案例如保罗·鲁道夫的曼哈顿下城高速公路计划。巨构的悖论在于用要素片段的抽象来指代无定形的基础设施网络，图像的可读可译功能变成了巨构的构成标准。1990年代开始，新崛起的景观都市主义试图整合不同的空间元素，构成整体连绵不断、局部亲切宜人的空间。弗兰姆普敦将这类空间整合策略称为巨形，以区别于前一代的巨构。巨形躲藏在城市的正常空间秩序之后，在亲和的外表下容纳多种灵活多变的功能与设施。构成巨形的逻辑不再是重力与易读性，而是各种物质与能量的运作机制与管理程序。

从某种“机会主义”的视角来看，桥梁是由翘曲的和折叠的结构体量组成的，这些异形的表面构成了许多难以设想的非正规空间，类似骨骼的结构系统具有同构型重复的特性，每个结构“晶体”单元可以成为安放居住模块的场所。因此，“桥舍”计划产出的四个方案都试图向桥梁结构的空隙塞入所需的居住功能。但是这仅仅是“桥舍”的构成逻辑的一小部分。在可见的结构之后，桥舍更关注运作一个城市功能所需要的物质、能量与交通条件。城市桥梁可以是“流动人口”的家园，也可以是新的城市生活的载体。流动人口通过收集他们自己的能源和降低对私有化的基础设施的依赖，形成了城市流通网络的寄生体（或共生体）。同时，桥梁作为一个畅通无阻的公共空间发挥其作用，它不仅代表着参与本课题的不同研究小组共同的研究兴趣，也回应了当前的城市弊病、“进入城市的权力”以及构建中国当代城市景观的要求。

## “桥舍”方案

### 方案1. 倒置之桥

方案希望探讨承担交通和生活综合体功能的基础设施新类型。从居住品质和普适化角度出发，选取最常见的中承式拱桥为原型，作为建筑体量的轻质钢结构体系和桥面一同悬挂于桁架拱之下。利用拱本身的结构空腔作为交通要道，到达建筑不同标高，再通过悬挂至桥面的垂直交通核连接各层。可抽拉的模块化单体提供预制装配的便利。悬吊结构赋予建筑自上而下生长的可能，城市天际线被倒置带来特殊空间体验。作为“经过”的桥和作为“停留”的宅赋予彼此新的交通流线和功能，并通过统一的结构合二为一。

### 方案2. 花市之桥

“桥舍”是对基础设施功能的复合利用，在通行之上叠加居住空间、景观空间、公共活动空间，由此，在节约土地资源的同时，解决部分外来流动人口的居住问题，回应城市中现实的社会问题。立体分离过境交通和慢行交通之后，在三跨的波形中承式拱桥之上用模块化的集装箱单元来组合搭建一个停车与居住相结合的生活社区。居住者可以在桥舍上停放自己的移动餐车，夜晚停车居住，白天出车工作，停车场转变为交流休憩空间。为了将商业商务中心和密集的居住区联系起来，在滨河景观带之上布置移动餐车停车点。设计希望能给居住在这个社区的低收入人群提供一个安全、私密的居住空间，也期望能给他们提供新的生活工作模式。

方案 3　物流之桥

针对如何有效合理地复合“桥舍”的交通和居住功能这一问题，方案从给不同的空间进行速度分级出发，试图建立一个人车交互的社区。不同速度的空间可分为：用于交通的快速道路空间，用于建立联系的慢速道路空间，用于居住的社区空间。之后在桥梁中间设置绿轴作为连接两岸的步行系统，以可移动的扩展单元联系绿轴上的居住区和商业区；同时，可移动单元也可用于运输，并让营业结束后的商户能回到居住区，在家里进行加工，获得归属感。通过这一系列的努力，希望基地较大的交通流量能给“桥舍”带来更多的发展机会。

方案 4　地形之桥

“桥梁”和“居住”两种功能在叠合之后，如何激发出一种全新类型？方案由基地出发，在通行和居住以及配套设施的功能之外，加入了景观的组织，而这些都在统一的桥梁的结构之上来实现：居住与商业附着于混凝土拱之上并随着曲线而流动，步道则延续到桥面和拱面的上下；而另一方面，系杆拱桥的基本形态也因其他功能的加入而进行了一定的变换。由此，结构作为承载，景观作为联系，所有的功能都从这样的扭合之中获得了自身新的特征。居住者体验的是一种全新的居住，行人使用的也是一座不寻常的桥，桥因舍而特殊，舍亦因桥而非常。

参考文献


[1] 鲁安东，窦平平 . 环境作用理论及几个关键词刍议 [J]. 时代建筑，2018（3）：6-12.

[2] Angus MacDonald. Structural Design for Architecture [M]. Woburn: Architectural Press, 1997.

[3] Alex Wall. Programming the Urban Surface [M]// James Corner, ed. Recovering Landscape, Essays in Contemporary Landscape Architecture. New York: Princeton Architectural Press, 1999: 233-249.

[4] 李麟学 . 知识、话语、范式——能量与热力学建筑的历史图景及当代前沿 [J]. 时代建筑，2015(2)：10-16.

[5] Frampton, Kenneth. Megaform as Urban Landscape [M]. Urbana Champaign: University of Illinois at Urbana Champaign, 2010.

图 3 保罗·鲁道夫的曼哈顿下城高速公路计划（Lower Manhattan Expressway）

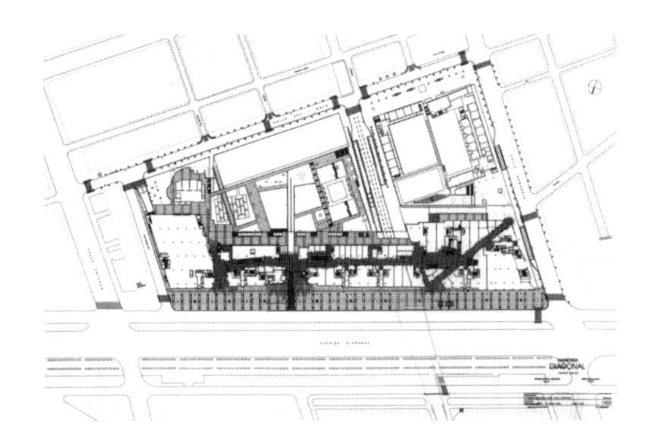

图 4 莫内欧与莫拉雷斯（Manuel de Solà-Morales）设计的 L'illa Diagonal 综合体

# Knowledge-based Design Education: An Experiment about Environment, Structure and Topography

TAN Zheng

**On the Knowledge of "Matters" and "Flow"**

Architecture is a discipline about how human beings intervene and reconstruct the material world. Environment, structure, and topography are the fields directly connected with the material world in the architecture context. "Environment" focuses on the intermediating function of the relationship between the building coordinator and the nature (Lu Andong, Dou Pingping, 2018); "Structure" focuses on the integration function of building balance form and content relationship (Angus MacDonald, 1997); "Topography" focuses on the connection of the building as a whole surface component (Alex Wall,1999). Broadly speaking, these three areas tend to boil down to "environment", which can cover all the knowledge associated with management and guidance of the spatial experience.

In addition to the knowledge of political and cultural expression, the knowledge of architecture will not be parted from the above three fields. However, in the current architecture education, the knowledge of environment, structure, and topography belongs to different course categories (building structure, building technology, environmental engineering, site design, landscape science, etc.), and it is not carried out organically in the whole process of the single design training. Students also lack the opportunity to apply and test various types of knowledge about the workings of a substance in a design. Based on this, it is urgent to construct a practical training course focusing on the "environment, " considering research and design as the core, so as to make up for the deficiency of the existing training system.

The teaching group led by Yung Ho Chang and Zheng Tan began the course experiment of "infrastructure architecture" in 2015, pinpointing the design to the broader urban infrastructure (or "structures") other than the traditional perception of buildings, students were encouraged to learn a avariety of matter, energy and structure knowledge to cope with the endless challenges of design in the real world. Based on the research-oriented design courses of senior undergraduates or graduate students, the project organized "Infrastructure Architecture" (2015) successively, "Waterfront Infrastructure"(2017) and "Qingpu Bridge" (2018) and carried out a series of complementary international seminars, symposiums and field surveys to envision possible future forms of energy, mobility and habitat conditions.

### Infrastructure and Environment

The modern infrastructure network has formed a new artificial terrain on the surface of the earth, and its significance should be re-examined from the perspectives of architecture, anthropology, and geology. The infrastructure is highly developed today, and the habitat environment has become increasingly subject to the control of various technical conditions. Infrastructure determines the measurement of people's spatial experience-thermal comfort, satisfaction, mobility, accessibility, security, and so on. This experience of "modernity" is increasingly parted from the inertia of culture and custom and becomes a universal requirement. However, the keys solve the housing crisis, poverty and social mobility — such as capital, labor, energy and all natural resources — remain in the hands of various types of "Interest-related parties", i.e. stakeholders. Therefore, the relatively abstract "infrastructure" is composed of specific systems, networks, mechanisms, institutions, and structures. Architecture is involved in infrastructure not only by understanding its technical operation rules but also by understanding its spatial organization and the game of various institutional forces behind it.

Infrastructure is an artificial structure system which allocates information, energy, transportation, and other public resources. At the macro-scale, it can be expressed as a variety of networks and systems, at the micro-scale, it is represented as a functional component or a public space which is closely related to daily life. The contemporary infrastructure is actively involved in the form of modern life, and the traditional streets, squares, and markets are integrated into the new public space. The transportation infrastructure space represented by various channels, hubs, and vans has become a place of collective memory, and the virtual public space composed of information and cultural infrastructure is also emerging, the mechanical features of infrastructure has begun to soften. It constitutes Benjamin's "internal" that cannot be escaped. And the infrastructure has begun to get rid of the cold, hidden image, becoming key elements of the impact of meaning, symbols and emotional occurrence in social life. Correspondingly, the knowledge system of "matter" and "flow" should be digested and translated, from the traditional pure engineering knowledge to the basic cultivation base of space planning and design.

### Bridge Shelter Design of Experiments

"Bridge shelter" graduation design course began in the spring 2018, it is a continuation of a series of courses on "infrastructure architecture", and is jointly carried out by Tongji University and Columbia University. "Infrastructure architecture" attempts to reset the consumption and multiplication of all factors of production and explore the design of a higher standard of dwelling forms on the basis of the infrastructure provided by the city, to partially achieve the autonomous sustainable use of energy and labor. These infrastructures include transportation, energy, logistics and other systems. Based on this, Tongji University and Columbia have run two parallel design teaching programs in Shanghai and Los Angeles to test- ① Planning of the mobile population housing project; ② Innovative urban housing forms and; ③ corresponding infrastructure solutions. In the specific teaching plan design, the California teaching program, which is based on the ongoing urban rescue plan, is centered on the transformation of the city's existing old urban structures (warehouses, garages, etc.) into homes for the homeless to return to society, while focusing on the eventual completion of the application of energy independence, financial balance housing combination and modular prefabricated assembly building technology. The Shanghai teaching program considers the combination of the housing of the low-incomers with a new urban bridge, at the same time, various energy-saving and green building technologies should be applied, the final built environment should also be integrated with the urban landscape system.

"Bridge shelter" is located in Qingpu District, Qingpu Avenue, Shanghai. The main body is the bridge across the Dianpu Lake, connecting Dianshan Lake Avenue and Huqingping Highway, on the Dianshan Lake Avenue, there is "Dianshan Lake Avenue" station of Shanghai Rail Transit 17 Line. Professor Yung Ho Chang of the teaching team has designed a road bridge that completely separates pedestrian from traffic, Which is now in the construction process. Inspired by the form of the bridge, the teaching team interpreted the bridge's mission as a "bridge shelter" project for the low-incomers. Bridge shelter first needs to meet the established bridge standards and appropriately higher than the standards. And a mobile population station must be accommodated within the Waters Blue Line to rehouse100 prefabricated assembly units for low-income people to work and live in. Bridge shelter's energy should be partially autonomous and provide the working space for low-income people to realize the self-sufficiency of the community energy supply and the reproduction of human value.

This subject requires losts of considerations:

1) The main body of the project is a bridge for the traffic between motor, vehicles, and pedestrians. Because the river is a VI - class passageway, the navigable clearance of the bridge is as follows: the net width should not be less than 30 meters, upper and bottom width not less than 22 meters, net height not less than 4.5 meters.

2) The original planed Qingpu Avenue road red line width is 50 meters, due to the realization of speed branch, without affecting functionality, the actual bridge width can be significantly narrowed. The project requires the bridge motor vehicle, non-motor vehicle and pedestrian system to connect well with the surrounding environment, and effective segregation between different modes of transportation. The surrounding area of "Dianshan Lake Avenue" station is conceived as the new city deputy center of Qingpu District. The "bridge shelter" should consider the chronic transport system and the long-term connection with the deputy center area of "Dianshan Lake Avenue." The possible case can be reasonably imagined for planning new means of transport with the system.

3) The project must be capable of accommodating 100 temporary resettlement stations for low-income mobile population and these stations must also be made to organize the production of these residents to achieve self-sufficiency in life. The total resettlement area of the station should be no more than 1,500 square meters, the living form of the mobile population can use miniature prefabricated living units (so-called capsules), which must contain the necessary living, sanitary, dining, communication and management facilities. The entrance, transportation, and logistics of the temporary housing should be separated from the daily traffic of the bridge.

4) The energy consumption of the project must be as self-sufficient as possible. Solar photovoltaic technology and related energy-saving technology for energy supply, collecting rainwater and recycled domestic water for landscape irrigation should all be considered.

And the project's landscape system must blend with the surrounding landscape. Based on technical support (structural evaluation and software optimization), the basic structural calculation of bridge structure safety is considered.

This program is driven by design knowledge, to achieve this, a "reading club" form was applied in the teaching process. This teaching form is usually used only at graduate stage. Students were asked to read professional works on solar architecture, prefabricated assembly technology, road traffic design and residential design, and then write and discuss book reviews. At the same time, professionals in the field of environment, construction and architecture were invited to participate in teaching to enrich the design knowledge and skills required by the group. The team organized a joint work camp with Columbia University in New York and an informal collaboration with the Texas Tech University, before the mid-term report and final report of the course.

The conclusions and reflections of the topic are as follows:

There is no lack of precedent in the history of modern architecture for the space type that is similar to "bridge shelter", throughout the 1950s and 1970s, the prevailing megastructure movement advocated the idea of erecting cities on infrastructure like highways, typically as the Lower Manhattan Expressway of Paul Rudolph. The paradox of the megastructure is that it refered to an amorphous infrastructure network by abstracting element fragments, the readable and translatable function of the image constituted the standard of the composition of the megastructure. From the 1990s, the new landscape urbanism tried to integrate different spatial elements to form a continuous on the whole, locally friendly and pleasant space. Frampton called such spatial integration strategy megaform to distinguish it from the previous generation's megastructure (Frampton, 2010). Hiding in the city's normal space order, megaform accommodates a variety of flexible functions and facilities under the appearance of affinity. The logic that makes up the megaform is no longer the gravity and legibility, but the operating mechanism and management procedure of various substances and energy.

From a certain "opportunistic" point of view, the bridge is composed of warping and folding structure volume, these irregular surfaces constitute a lot of informal spaces that are difficult to conceive, which are similar to the repetitive structures of the skeleton system, and each "crystal" unit structure can be used to place living modules. Therefore, the four proposals of the bridge shelter project all attempt to fill the gaps in the bridge structure with the required residential functions. But this is only a small part of the logic of the "bridge shelter." Behind the visible structure, the bridge shelter pays more attention to the material, energy and traffic conditions needed for a functional city. The urban bridge can be the home of "the mobile population" or the carrier of new urban life. By collecting their own energy and reducing their reliance on privatized infrastructure, mobile populations form a parasitic (or symbiotic) network of urban distribution networks. At the same time, the bridge functions as an unimpeded public space, it does not only represent research interests shared by different research groups involved in the project but also responds to the current urban illness, "rights to the city," as well as the requirements of the construction of Chinese contemporary urban landscape.

**"Bridge Shelter" Schemes**

Scheme 1. Bridge as Inverted Skyline

The infrastructure serving for transportation and dwelling is discussed in our project as a new typology of space. From the perspective of living quality and universalization of construction, we picked the half-through arch bridge as a prototype. Due to the feature of the half-through arch bridge, the light steel structure system and the bridge deck can be suspended under the truss arch as a whole. Regarding the circulation, cavities of two arches are used as main lanes to reach different levels of the building. Vertical traffic cores are added to connect each floor to the bridge deck. Sliding dwelling modules are available for prefabrication and assembly. In addition, the suspended structure provides the opportunity of growing from top to bottom and demonstrates a scene of the inverted skyline. The bridge for traveling through and the dwelling for lingering in form a united system, which would upgrade their existing function and social form.

Scheme 2. Bridge as Flower Market

"Bridge shelter" is a device about composite utilization of infrastructure to serve multiple needs such as mobility, dwelling, communication, retailing and recreation. While saving land resources, it can solve the housing problems for migrant population and respond to the social problems in the city. After separating fast traffic and slow traffic in the three-dimensional way, modular container units are employed to build a living community on the three-span half-through arch bridge. Residents can park their own dining cars on a common deck so that the spared space can be used as a car park and dorms at night, and shops in the daytime. In this way, the parking lot is transformed into a congregational space. In order to connect the flea market and residential area, the car park is arranged for dining cars along the riverside. We hope to provide low-income people with safe and private living space as well as a new life style.

Scheme 3. Bridge as Logistic Hub

As a response to the problem of how to integrate the functions of traffic and dwelling in a smart way, this project starts from speed filtering for different spaces and tries to establish a community where people and vehicles interact with each other. Spaces for different speeds can be sorted into different types, including high-speed motorways for traffic, slow non-motorized roads for establishing connection and residential elements for living. A green belt is paved in the middle of the bridge as pedestrian system to connect the two riverbanks. The users of movable modular units and visitors of passing by the public spaces are protected from intervening into each other. At the same time, the mobile units can be used for transportation as well, and merchants can return to the residential area by using the mobile unit at the end of the business day and process their merchandise at home which certainly will help them gain a sense of belonging. By focusing on the large-scale traffic flow at the site, we hope to bring more development opportunities to the "bridge shelter."

Scheme 4. Bridge as Land Form

The topic of "bridge shelter" raises a question about how to inspire the potential of a new prototype after the coupling of the functions of the bridge and dwelling. Given the condition of the site, the organization of landscape is introduced to the functions of transport, dwelling and supporting amenities. All programs will be arranged upon a main structure of the bridge. Dwelling and amenities are attached to the concrete arches and cascade along the curvature. A landscaped footbridge undulates above and below the driveway and arches. In addition, the general form of the bowstring arch bridge is shaped as these functions join together. Thus, when the structure and landscape perform their independent roles, all the functions have regained their new features through the posture of twisting. Space users are enabled to enjoy a new life style while the pedestrians are using an atypical bridge. The shelter and the bridge become special because of each other.

# 以热力学为线索的设计方法论：关于哈佛设计研究生院热力学设计课程的过程与思考

陈昊、胡琛琛

## 课题与设计过程

哈佛设计研究生院自2012年起在自选设计课环节开始出现以热力学为研究方向的课题，持续四年；课题结合了高层综合体、城市更新、基础设施等多个命题。其间，学院设置了诸多与之配套的讲座与研讨课，邀请马提亚斯·舒勒、萨尔曼·克雷格等环境工程和可持续领域的科学家、顶尖工程师，分享实践经验和正在进行中的研究。

其中，2014年春季课是唯一一次选址中国的课题，题为“热力学唯物论在高密度都市聚合体中的运用”[1]。课题关注中国城市化新浪潮中出现的中型城市：它们多以大型基础设施为特色，追求可持续发展和生活品质的提高。义乌和青岛是两座中型城市，具有不同的地方特点与气候特征，但同样计划依托高铁站来展开新一轮城市开发。课题试图从类型学和热力学的角度重新思考中国城市发展中的三个事物：高速铁路站、CBD、巨型街区。

我们试图通过研究非常规的建筑原型，将三个看似分离的要素结合起来，探索混合使用聚合体在提高公共生活品质和建立新的设计方法上的潜在可能。将形式、能量、物质、身体关联成一体——可称之为一种新的“热力学唯物论”。

热力学唯物论聚焦于设计过程并回归到一种建筑的整体观，即将可见或不可见的材料性，通过最大程度的融合，构成我们对于建筑和城市的体验。形式、身体、自然因素、材料、功能、时间与美都是交织构成“材料性”的基本分类。在建筑语境中，它表示某个时代的物质文化、建造材料以及形态，作为决定建筑体验和完成效果的综合因素。它既是客观的，也是主观的；既是个体的，也是集体的。

学生可以从两个城市中选择一个作为课题的基地。和青岛的场地相比，义乌的基地在自然地形和城市文脉上更具体和丰富一些，潮湿少风的气候也较青岛更容易挖掘特点，利于在不同层面上发展出多重的设计方向。对比往年，2012、2013年课题所选基地为巴塞罗那和马德里，其大陆性气候特点非常鲜明，而义乌相对温和模糊。那么，在暖房和遮阳棚的终极类型[2]之外，亚热带季风气候是否存在某些范式？通过热力学得到的形式是否会颠覆我们的过往经验？

1. 风

风如何限定建筑的选址、朝向和体量？

通过对义乌各项气候参数的分析，可以发现：终年少风是义乌气候最显著的特点，风速均在3~4kts之间（约为1.5~2m/s），以东北风为

---

1. 哈佛设计学院2014春季课程手册，第61页。
2. 伊纳吉·阿巴罗斯：《建筑热力学与美》，同济大学出版社，2015年，第24页。

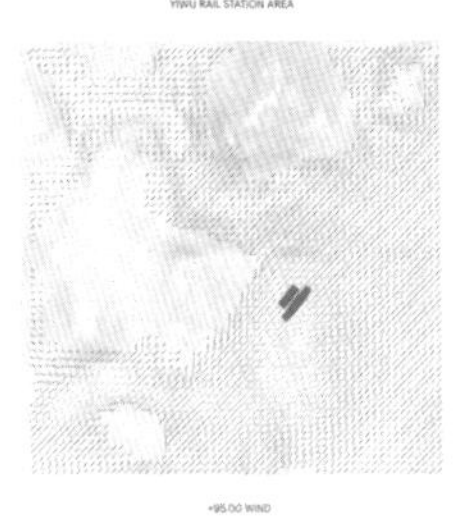


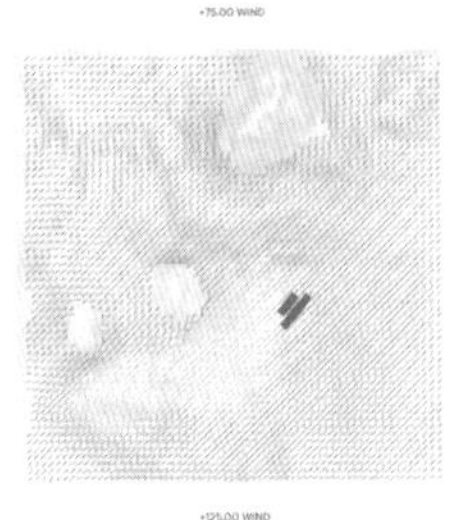


图 1　受地形影响，125 米标高处的山脚附近呈现出最显著的通风走廊

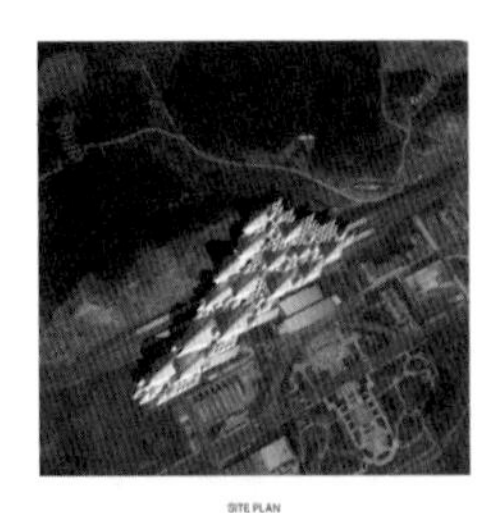

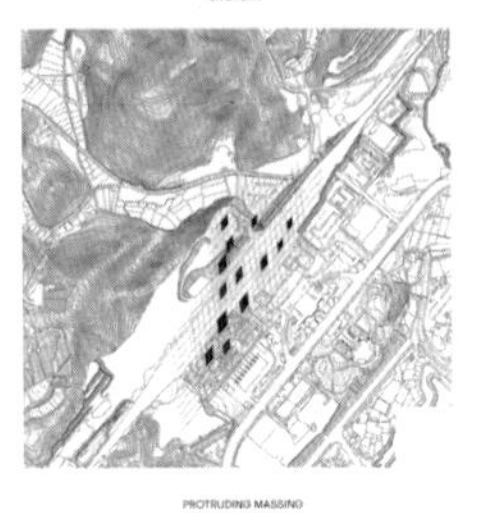


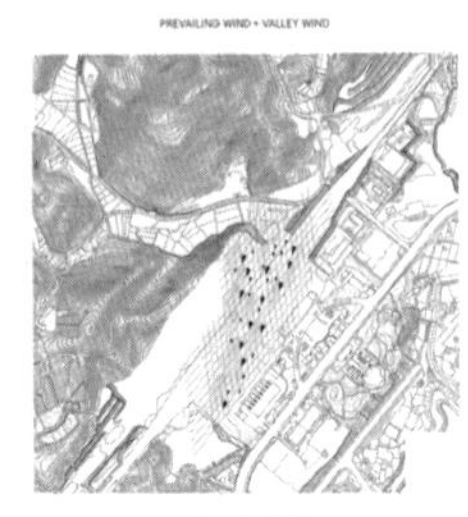


图 2　两个方向的主导风向、顶部凸起的烟囱体量、屋面采光通风的洞口构成了建筑的整体轮廓

主导风向。结合焓湿图可见，在义乌利用主导风向，加强建筑内的空气对流，是提高体感舒适度最有效的策略。

观察基地地形又能衍生出进一步的发现。义乌高铁线东南面向城市，西北靠山。从大区域来看，高铁站及其西侧的平坦区域被西北最高的山峰和东南稍缓的丘陵夹住，在此形成加强东南风速的通风走廊，其加速在海拔 125 米时最为显著。从小区域微气候来看，如果考虑在高铁站北侧公路周边密植乔木，降低北侧温度，形成南北温差，再形成由北向南的凉爽气流，有利夏日通风降温（图 1）。

将加强空气对流的方式缩小到建筑尺度来看，最常用的通风形式不外乎两种：空气的水平对流，即穿堂风；空气的垂直升腾，即烟囱效应。前者通过限定建筑体量的间隙（用于外部空间通风）和立面的开洞位置（用于室内通风）来获得；后者通过加热低处的空气使其上升，并吸入底部的冷空气来获得。义乌夏至日太阳高度角 83.1°，冬至日太阳高度角 40.1°。那么，义乌的理想烟囱模型或许是：南立面与地面夹角 83.1°，夏季形成自遮阳；北立面作为烟囱受热表面与地面夹角 40.1°，尽量多地捕捉全年日照。

通过以上分析可以大致勾勒出建筑粗浅的轮廓。它的网格由两套风向叠加而成：全年主导的东北风，和微气候形成的北风。竖向上建筑形体布置在风速最大的海拔 125 米（即离场地标高 8.5 米）最理想。如果希望进一步加强空气流通，根据建筑网格设置一系列倾斜的“烟囱”会起到显著的拔风效果，其倾斜角也会起到均匀南北立面温差的作用。同时，“烟囱”集群在屋面上形成狭窄密集的通风走廊，可有效降低夏季的屋面温度。烟囱的分布方式由东北向西南逐渐加密，以减少对东北风的遮挡。屋面其他区域也被更小尺度的凸起所覆盖，作为无烟囱功能的采光天窗，构成了屋面整体密集凸起的统一肌理（图 2，图 3）。

2. 功能

建筑如何被理解为第二自然？最初的建筑不

就是人在岩洞中的栖居吗?

如果说风定义了建筑的总体轮廓和能量利用策略，那么功能的加入会进一步细化能量的利用方式，同时，能量的分布和利用方式也会反过来引导功能布局。

这里所说的功能不只是一份须被满足的房间列表，更多是关于当代基础设施使用方式和身体体验的构想。高铁车站内能否实现更接近自然的舒适体感? 候车大厅如何成为人、物资、空气、光线共同流动的平台? 建筑形式与空间如何黏合关联起各种形式? 这些问题或许并无新意。但热力学给这些问题提供了新的支点，试图将它们整合进同一个能量系统，或将得到不同以往的空间类型与体验。

综合之前的分析，通风需求最迫切的候车大厅被布置在 8.5 米标高，形成漂浮于高铁线上没有气候边界的大平台，自然风从中穿梭而过。楼板分布着大小各异的孔洞将人流和自然光引向下方的站台，将冷空气从底部吸入；顶板也分布着联通烟囱的孔洞，将热空气通过烟囱排出。等候室、商店、票务等封闭且带有空调的功能单元分散在半室外的大平台上，既保证了停留区最佳的体感舒适度，又能有效减少能耗，并且调节了大型基础设施不够友好的空间尺度(图 4—图 6)。

车站顶部若干烟囱塔楼提供了小型办公与商务酒店，服务于往来义乌的经贸人士。剖面上看，各楼层通过墙顶设置侧缝或进行错层设计，向烟囱内排出热空气，从外立面低处取风引入冷空气(图 7，图 8)。

3. 立面与材料

塔楼立面总体采用遮阳百叶包裹，在不同朝向上有所变异。从太阳辐射模拟来看，由于形体倾斜的原因，各立面辐射量与常规的立方体稍有不同。在夏季，南立面的辐射量相对较小，西立面较东立面辐射大，北立面与东立面接近；在冬季，各立面辐射量比较均匀。由于形体已对南面产生自遮阳效果，南立面反常规地采用全透明、无外遮阳的做法，利于冬日辐射采暖；东西立面考虑遮阳需求，采用东疏西密的横向百叶；北立面采用竖向百叶，形式逻辑大于实用意义。相同垂直间距的水平百叶在曲形立面上会形成下宽上窄的遮阳深度，很自然地适应了不同倾角立面的遮阳需求。烟囱外壁的两个面采用透明玻璃，内壁采用利于吸热的深色混凝土，使阳光能够透过外壁照射到内壁上，加热烟囱内的空气。

这些热空气不但来自于塔楼本身，也来自于下方开敞的候车平台；塔楼的存在不单是对车站功能上的补充，也是下方公共空间的拔风通道。整个热力学系统或许就是义乌气候条件下城市基础设施聚合体的一种可能原型。

## 热力学作为探索形式的方法论

GSD 所推动的热力学研究本质上是一种形式生产的方法论，试图将气候、能量、材料等都融合成为推动形式产生的要素。

尽管它常被误解为高度依赖热工知识和参数化模拟，但实际上，它并非“唯技术论”的设计方法。做个不恰当的类比：在没有计算机辅助结构找形的 19 世纪末，高迪的圣家族教堂就无法被设计了吗? 热工知识是提出假想的基础，参数化模拟是验证假想的手段，而设计思维是整合这些技术的网络。通过建筑师的判断、筛选、调整，技术数据才得以转化为空间与建构。整个设计并非是线性进化的过程，而是往复动态的；而设计成果也是对“事实”出于不同目的或“意图”的选择性结果。因而，即使在同一基地里研究同一组数据，由于个体主观有所差异的解读和转译，最终也会呈现出不尽相同的建筑。

可见，与常规的建筑学设计训练相比，热力学的研究框架更接近设计思维训练，呈现的结果也更像是用来支撑论点的一系列图解。如前文所说，其本质是探索形式的方法论，也是诸多形式

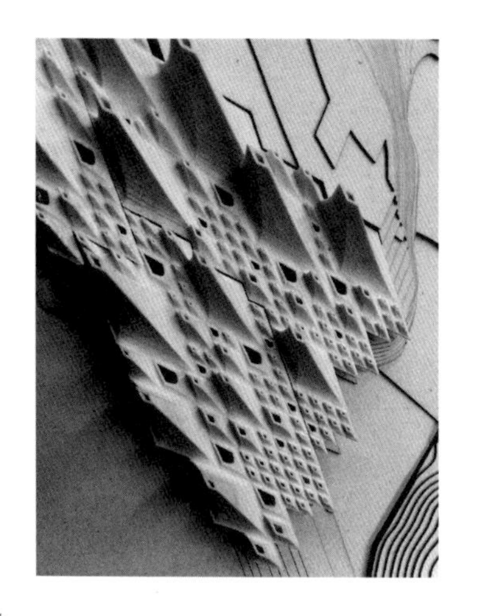

图 3 实体模型照片

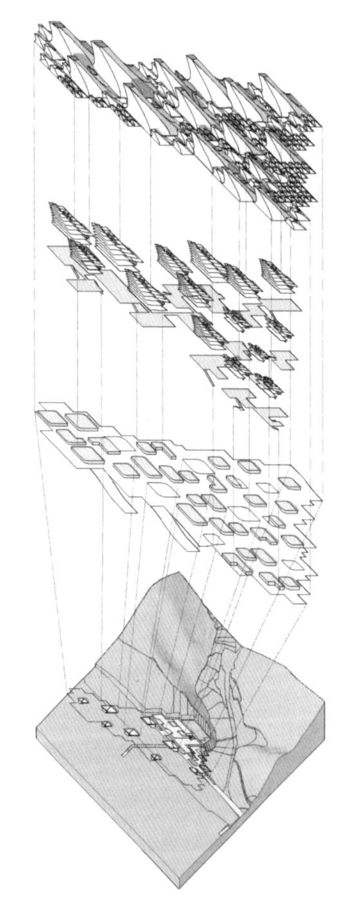

图 4 分解轴测示意。由下至上为：场地地形与站台层，候车开放平台与分散空调空间，烟囱内的居住与办公空间，布满凸起和洞口的屋面

产生方法论中的一种。在 GSD 百家争鸣的大环境下，它与其他设计课程相比并不新奇，只是研究对象和研究工具有所不同。[3]

这些课题的共同特点是强调设计的意图和方法论的演绎，弱化使用功能和建造。简单来说，就是在方法论的框架下，通过图像与模型将论点最大程度展示得清晰和完整。在 GSD 的评价体系中，能在各个设计阶段贯彻课题设定的方法论，要比抛出优美的方案更为重要。这种方式虽然看似不鼓励创造，但大大加强了设计的可教学性，让学生得以掌握在日后实践中可被重复利用的思考模式，或是以相似的框架构建自己的体系。

当然，热力学唯物论与 GSD 其他设计课在方法论的具体内容上还是存在较大差异。它不囿于纯粹的建筑学理论，试图引入自然条件和技术要素，拓展设计思维的边界。热力学相当于给建筑设计提供了一个可在不同尺度与层面上细分延展的框架：每个子项的每个层次都能推导出多种可能，设计者可一边通过不断挖掘下一层级来纵向丰富设计，一边尝试各子项间不同的排列组合来横向比较推敲。这种网状的、扁平化的思考方式，与由“前期研究—建立观点—获得形式—具体深化”的线性思考方式带给设计者的感受是相当不同的。前者常困扰于如何横向筛选，而后者常困扰于如何纵向深化；前者困扰于寻找形式，而后者困扰于方法论得到的形态或许有悖主观审美。这也许是阿巴罗斯教授一直把热力学原型称为“怪物”的原因：它们常常看起来古怪而有悖常识。

## 热力学建筑的局限性

### 1. 教学中的局限性

热力学建筑作为理想模型是有效而且有趣的，但如果试图将它普及到职业教育中，则存在一定的困难。因为热力学建筑的讨论主要基于热力学的理想模型，以能量和温度作为推敲空间与形式的切入点；而在建筑所要满足的诸多需求之中，能量的流动尽管必不可少，然而也仅仅是其中一部分。尽管热力学试图去容纳建筑学的方方面面，但似乎仍有待克服技术与建筑学由来已久的隔阂。

这种隔阂究其产生的原因：其一，是建筑学科对建筑技术缺乏重视，国内建筑学院的基础教育核心是围绕空间形式展开的，建筑学院很难将技术学科归于其中。一方面，主流建筑教育对空间形式的探讨主要基于功能布局、结构形式与材料构造展开，往往忽略了热力学理论与建筑空间形式的有趣关系。并且通常，即使建筑中出现某些热力学原型如中庭、烟囱、通道等，也往往是出于功能、流线等因素的考虑，而非单纯热力学本身。另一方面，暖通、电气等技术学科本身就很庞大，建筑类的应用只是几大方向的其中一个分支，要促进学科融合还需要环境工程师对建筑学有相当程度的理解，并在课程设置和教学人员配置上有所交叉。在教育过程中如果缺乏对暖通等建筑技术学科的基础知识体系的了解，将会导致对热力学设计缺乏理论基础。 所以阿巴罗斯教授在前两年的热力学设计课程中，花费了整个学期近一半的时间让学生进行热力学基础知识的探索研究训练，并邀请马提亚斯 · 舒勒等建筑技术领域的专家进行原理知识授课。

其二，是热力学的形式导向本身存在的局限性。从若干年设计课程的过程和成果综合来看，不难发现很多热力学原型都是将某种被动式生态策略超尺度放大，如水平或垂直风道、上大下小产生自遮阳退台、根据阳光和风向切割的体量。这种图解式的形式策略是相当概念化的，弱化了功能、结构、建造等诸多现实因素。你可以说它挑战了思考建筑的传统范式，但也很难避免其形式与常规的建筑学的教学目标产生矛盾。

2. 实践中的局限性

一般情况下，境外设计机构对于有生态设计要求的项目在初始阶段就会邀请生态顾问参与。而国内的设计项目鲜有业主或建筑师在概念阶段就将生态设计作为整体的一部分，而后期介入的暖通工程师也相对被动，只是配合建筑落实基本的暖通设备图纸，难以对建筑空间有深入的认识，更无法在方案伊始就以积极的态度通过热力学对形式与空间提出启发性建议。因而在实践中，尽量减少设备对空间和形式的影响，尽管消极但已是万幸。

另一方面，即使热力学可以在概念方案伊始提供导则性的设计策略，这种理想模型也常会被建筑学和建筑实践的其他诉求以各种方式消解。有别于设计教学，设计实践无法逃避使用需求和建造成本，设计的过程是反复平衡各方矛盾的过程。通常情况下，基本需求要优先于期望需求被满足。热力学建筑所提倡的被动式提高体感舒适度，在一般民用建筑中并非刚需。即使某些中庭或是天井在生态方面起到了一定作用，其设计初衷往往是优化空间和利于疏散。

## 热力学设计的改进

热力学建筑是建筑技术学科的一部分，并非崭新的学科。在过去五年内，哈佛设计研究生院将热力学置于学院教学大纲的重要位置，可见建筑设计正试图回应当今社会发展有关能源的问题与挑战，并唤起教育界对建筑技术学科的重视。

1. 热力学设计应该如何拓展?

首先，从观念的普及入手，从职业教育开始向建筑师输送更全局的热力学观念。提到热力

—

3. 如“Ornament Space”（Chris Kerez）从图像学出发，将平面形式母题在空间中重构、再演绎；“Cross Border City”(Chris Lee) 从类型学出发，将城市中既有深层结构转化为新的城市策略；“La Strada Novissma ”(Johnson Mark Lee) 从类型学出发研究城市街道中半自治的独栋建筑。

图 5 候车平台

图 6 站台层

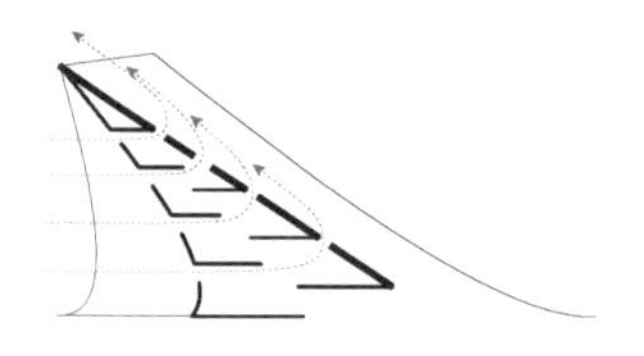

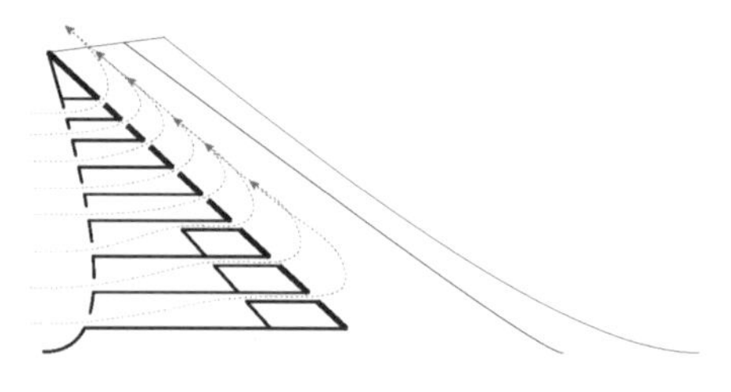

图 7 大小不同空间中烟囱内部气流示意

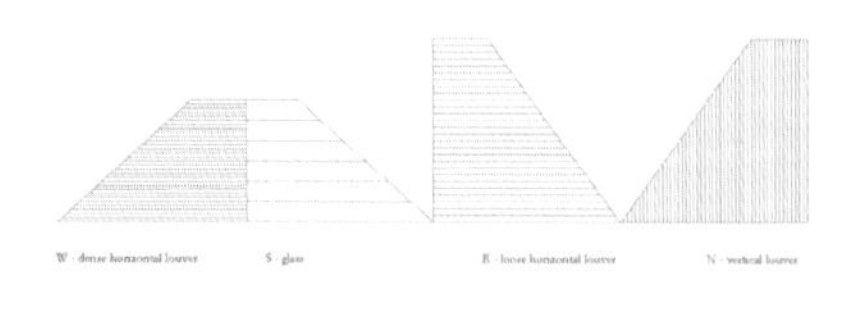


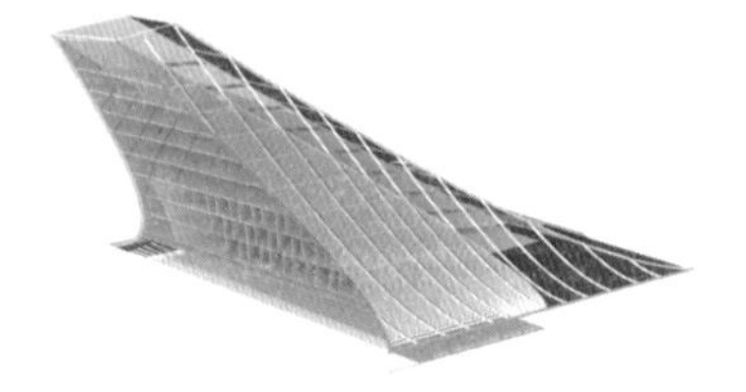

图 8 各朝向立面回应气候条件的形式构成

学建筑设计，很多人会联想到绿色设计、生态设计、整体设计、可持续设计等不同概念，或是BREEAM、LEED、绿色建筑认证等各类标准。尽管它们常常被学者提及和热论，却更像一种滞后的解决问题方式和一份不得不完成的附加列表，而非设计过程中的必要线索。

与之相比，结构与建筑的结合似乎被挖掘得更多，并在建成效果上更引人注目。虽然热力学本身不像结构体系对于构建空间的作用那么根本和显著，但它的影响同样由来已久又无处不在，主动或被动地影响人的感受。我们是否也应该像将结构与空间建构紧密结合一样，将建筑技术同样紧密地纳入建筑学的学科基础教学呢？是否可能在技术课程中加入形式的探索，在设计课程中融入技术的要素？技术与建筑相融合的观念如果能在建筑学本科教育时期植入，或将缩小各工种在实践中的立场和观念的差异。

2. 热力学设计应该如何退回?

另一个问题是，如何在热力学的合理性之余寻求丰富性与意义？继续前文热力学与结构的类比，如果说做到结构表现主义相对是容易的，那热力学表现主义本身能否具有相似的韵律美呢？或者更进一步的追问是，热力学应该在多大的范围和深度上被表现呢？是该彰显它还是隐匿它？如果彰显它又该如何避免沦为装饰或符号？彰显热力学的代价从建筑全局来看是否仍然可持续？这都是需要因地制宜讨论的具体问题。将理想原型直接运用似乎将这些问题简单化了，在现实中未必是恰当的提案。

因此，热力学建筑的原型设计方法提供了庞大的菜单，如何在进退之间找到合适的位置、做出恰当的选择是新工具对建筑师经验和判断力提出的考验。热力学在实践中的运用需要退回到关于建筑本质的思考，在合理性和意义之间寻求平衡，或更进一步将合理性转化为空间意义（这同结构与建筑的关系是统一的）。热力学作为启发性的工具，其目的不是彰显形式本身，而是让形式深度融合成为可感知和揣摩的某种存在。这些关于氛围、空间质量和身体体验的思考在教学和理想模型中是很难体现的，但在建成环境中尤为重要。如何在创造美好的物理氛围（温度、风、光线）的同时，反思非物质的意义，也许会给热力学建筑设计方法带来进一步的转向与展开。[4]

参考文献

[1] 伊纳吉 · 阿巴罗斯. 建筑热力学与美[M]. 同济大学出版社，2015.

[2] Iñaki Ábalos. Why we don't draw arrows (hardly ever)[J]. AV Monographs 169, 2014.

[3] 李麟学. 知识 · 话语 · 范式——能量与热力学建筑的历史图景及当代前沿[J]. 时代建筑，2015(2).

[4] 威廉 · W. 布雷厄姆. 热力学叙事[J]. 张博远，译. 时代建筑, 2015(2).

---

4. 这一观点借鉴了坂本一成对于白之家的解读：白之家是筱原一男的一个人生选择，或者说他建筑观的选择。他在将成为一个结构表现主义者，沦为一个所谓的结构合理性追随者的时候，及时地踏出了一步，在合理性和非合理性之间进行了一个非常重要的刹车，或者说是做了一个转向。这也许就是他之后建筑能够展开的重要的一步。

# A Design Methodology with Thermodynamics as a Clue

**CHEN Hao, HU Chenchen**

**Project and Design Process**

It has been four years since 2012 for Harvard Graduate School of Design to take thermodynamics as a research direction in the course of option studio. This project combines many propositions like high-rise complex, urban renewal, and infrastructure. Meanwhile, GSD has set up many relevant lectures and seminars, inviting top engineers and scientists in the field of environmental engineering and sustainability including Mathias Schuler and Salmaan Craig, to share their experiences on the research they were conducting.

Among them, the spring course of 2014 is the only project located in China. The project is "Application of Thermodynamic Materialism in High-Density Urban Complex — Two Chinese Cases." It focuses on medium-sized cities in the new wave of urbanization in China: most of which are characterized by large infrastructure, aiming for sustainable development and improving quality of life. Yiwu and Qingdao are two medium-sized cities with different geographical and climate characteristics. In line with the building of high-speed rail stations, they are planning for a new round of urbanization. The project tries to rethink three things in China's urban development from the perspective of typology and thermodynamics: high-speed railway station, CBD, and huge blocks.

We try to combine these three seemingly separate elements through the study of unconventional architecture prototype, exploring the potential possibilities of improving the quality of public life and new design method by using building complex. We call it a new type of thermodynamic materialism that connects form, energy, substance, and body together.

Thermodynamic materialism focuses on the design process and returns to a holistic view of architecture, that is to say, through the greatest degree of integration of material visible or invisible, our experience of architecture and city will be constituted. Form, body, natural factors, material, function, time and beauty are basic categories composing material... In the context of the building, it represents a material culture, building material, and forms in some era, objective or subjective, individual or collective, it's a comprehensive factor of construction experience and complete finish effect.

Students can choose one of the two cities as the base for the project. Compared with Qingdao base, Yiwu is more specific and rich in regard ot natural topography and urban context, and with the humid climate with less wind, it is easier to tap the characteristics, beneficial to develop multiple design direction on different levels. Different from Barcelona and Madrid which were selected as the bases in 2012 and 2013 with their distinctive continental climate features Yiwu is mild and fuzzy. So, in addition to the ultimate type 2 of greenhouse and awning, is there any paradigm for subtropical monsoon climate? Will the form we get from thermodynamics overturn the form of our previous experience?

1. Wind

How does wind define the location, orientation and volume of the building?

Through the analysis of various climate parameters in Yiwu, we found a most prominent : throughout the year, there is little wind. The wind speed is between 3 and 4 kts (around 1.5~2m/s), with the northeast wind as the dominant direction. Using the psychrometric chart as a reference, we realized the most effective strategy to improve the physical comfort to make use of the dominant wind and enhance the air convection in the building.

Yiwu high-speed railway faces the city in the southeast and the mountain in the northwest. From the large area, the railway station and the flat area of its west side is infibulated by the highest mountain in the northwest and a gentle hill in the southeast, this contributes a ventilated corridor which strengthens the southeast wind speed, the most notable acceleration is at an altitude of 125 meters. From the small area and looking the microclimate, if we consider planting trees around the north side of the highway of the high-speed railway station, we can reduce the temperature of the north side and form the temperature difference between the north and the south, and then the cool air flow from the north to the south will be formed, which is conducive to summer ventilation and cooling (Fig.1).

From the perspective of reducing the air convection to the building scale, the most commonly used ventilation forms are two kinds: the horizontal convection of air, namely drafts; The vertical rise of the air, namely the chimney effect. The former is obtained by limiting the amount of space in the building volume (for external ventilation) and the opening position of the facade (for indoor ventilation), the latter is obtained by heating the lower air to make it rise, and inhaling the cold air at the bottom. The solar altitude angle of Yiwu summer solstice day is 83.1°, a winter solstice solar altitude angle is 40.1°. In that way, the ideal chimney model for Yiwu may be: the angle between the south facades and the ground is 83.1°, forming its own shade in summer; and the angle between the north facade as the chimney's heating surface and the ground is 40.1°, capturing year-round sunshine as much as possible.

The above analysis can outline the rough contour of the building. Its grid is superimposed by two sets of winds: the year-round dominate northeast wind, and the north wind caused by microclimate. The ideal vertical placement altitude where meets the maximum wind velocity is 125 meters above sea level (i.e. the elevation from the site is 8.5m). To further enhance the air circulation, a series of tilting "chimney", which will play a significant pull wind effect, can be set according to the construction grid, and its tilting angle will also play a role of balancing temperature difference between north and south facade. At the same time, "chimney" cluster on the roof will form a narrow and dense ventilation corridor, which can effectively reduce the summer roof temperature. The distribution of chimneys is becoming denser gradually from the northeast to the southwest, reducing the shielding of the northeast wind. Other areas of the roof are also covered by smaller bumps, which, as a lighting skylight with no chimney function, constitute a unified texture of the roof's overall dense bulge (Fig.2-3).

2. Function

How can architecture be understood as second nature? Isn't cave the original form of human dwelling?

If wind defines the overall outline of the building and the strategy of energy utilization, the additional functions will further refine the way how energy is used, and the distribution and utilization of energy will, in turn, lead to functional layouts.

The functions described here are not just a list of rooms to be planned, but more about the idea of using modern infrastructure and the physical experience. Can a more natural sense of comfort be achieved within the high-speed rail station? How can the waiting hall become a platform for people, materials, air, and light to flow together? How are architectural forms and spaces bound together in various forms? These questions may not be new. But thermodynamics provides a new fulcrum to them, trying to integrate them into the same energy system or to get different space types and experience.

Integrated with the previous analysis, waiting hall whose ventilation need is the most pressing was marked up to 8.5m high, forming a large platform without climatic boundary floating on the high-speed line, and the natural wind travels through it. The floors are distributed in different sizes of holes that bring people and natural light to the platform below and inhale the cold air from the bottom. The roof is also distributed within the hole connected with the chimney, and the hot air is discharged through the chimney. Waiting rooms, shops, tickets and other closed unites without air condition are dispersed in a large half-outdoor platform, which will not only ensure the best body comfort in the stay area, but also effectively reduce energy consumption, and adjust the spatial scale to make large infrastructure friendlier (Fig.4-6).

Many chimney towers at the top of the station offer small office and business hotels that serve those who travel to and from Yiwu. On the section, each floor shall discharge hot air into the chimney through the side seam or cross-layer design on the top of the wall and take the wind from the lower part of the outside facade to introduce cold air (Fig.7-8).

3. Facade and Material

The tower facade is generally covered by sunblind, which varies in orientations. From the solar radiation simulation, due to the tilting shape, the vertical radiation is slightly different from the regular cube. In summer, the radiation of the southern facade is relatively small, the west facade radiation is larger than that of east facade, the north facade radiation is close to that of the east facade; and in winter, the radiation quantity of each facade is relatively uniform. Because the shape has its own sun-shading effect to the south, the south facade unconventionally uses all transparent without external shading, which is conducive to the winter radiation heating. The east and west facades have considered shading demand, horizontal louver whose western shutters are denser than eastern ones; the north facade adopts vertical blinds, formal logic outweighs practical significance. The horizontal louvers with the same vertical spacing will form a wide-below-narrow shading depth on the curvy facade, which naturally adapts to the shading requirements of different inclined angles. The outer wall's two surfaces of the chimney use transparent glass, and inner wall adopts dark concrete which is apt to absorb heat, so that sunlight can illuminate the inner wall through the outer wall, and the air inside the chimney is heated.

The hot air comes not only from the tower itself but also from the opening waiting hall platform downstairs; the tower is not only a supplement to the function of the station but also a draft channel for the public space below. The whole thermodynamic system may be one of the possible prototypes of urban infrastructure complexes in Yiwu climatic conditions.

**Thermodynamics as A Methodology of Form Exploration**

GSD promoted thermodynamic research is essentially a form of production methodology that tries to integrate climate, energy, and materials into the elements of a driving form.

Although it is often misunderstood as parametric simulations, which is highly dependent on thermal knowledge, in fact, it is not a "technology-only" design method. Make an improper analogy: could Gaudi's Sagrada Familia not be designed at the end of the 19th century without computer-aided design? Thermal knowledge is a hypothetical foundation, parametric simulation is the means of verification, and design thinking is the network of integration of these technologies. Through the architect's judgment, selection and adjustment, the technical data can be transformed into space and construction. The whole design is not a process of linear evolution, but a reciprocating dynamic; and the design result is the selective result of "fact" for different purposes or "intention." Therefore, even in the same base, the same set of data, due to individual subjective differences in interpretation and translation, different buildings will eventually be presented.

It can be seen that, compared with the conventional architecture design training, thermodynamic research framework is closer to the design thinking training, and the result is more like a series of diagrams to support the argument. As mentioned in the preceding article, its essence is the methodology of form exploration, and it is also a type of methodology of many forms. In GSD, the great environment, it is not a novelty compared with other design courses, only the research object, and research tools are different.

The common characteristics of these projects are the emphasis on design intentions and methodological deduction, weakening use functions and construction. In simplification, it is in the framework of the methodology, through the image and model, to maximize the clearness and completeness of the extent of the argument. In GSD evaluation system, it is more important to implement the methodology set in each design stage than to throw a graceful plan. This approach, although seemingly discourages creation, greatly strengthens the instructional nature of design, allowing students to master the thinking patterns that can be reused in future practice, or to build their own systems in a similar framework.

Of course, thermodynamic materialism still has a big difference in the specific content of the methodology with GSD's other design courses. It is not confined to the pure architecture theory; it tries to introduce the natural condition and the technical factors, and to expand the design thought boundary. The thermodynamic equivalently provides a framework for architectural design that can be subdivided and extended at different scales and levels: multiple possibilities can be derived from each level of each sub item, the designer can continue to explore the next level to enrich the design vertically, while trying different permutation and combinations of the sub-items to compare horizontally (Fig.9-11). This net, flat way of thinking brings different feelings to the designer from the linear thinking model of "preliminary study-build viewpoint-obtain form-specify and deepen." The latter often puzzles how to deepen vertically, while the former is often plagued by horizontal screening; the former is troubled by the search for forms, and the latter is plagued that methodological forms may be contrary to subjective aesthetics. It may be the reason Professor Ábalos has been calling thermodynamics prototypes "Monster": They often look strange and contrary to common sense.

## Limitations of Thermodynamic Architecture

1. Limitations in Teaching

As an ideal model, thermodynamic architecture is effective and interesting, but there are some difficulties trying to popularize it to vocational education. Because the discussion of thermodynamic architecture is mainly on the ideal model, energy and temperature are used as the cut-in point for the deliberation of space and form; and the flow of energy, though necessary, is only a part of the many demands that the building is to satisfy. Although thermodynamics attempts to accommodate all aspects of architecture, it seems there is still a long-standing gap between technology and architecture.

The reason the gap exists may be: first, is insufficient attention to the architectural discipline, the elementary education core of domestic architectural institute is revolved around the spatial form, so the architectural institute is difficult to attribute technical discipline to it. On the one hand, the discussion on spatial form of mainstream architecture education is mainly on function layout; structure form and material structure, people often neglect the interesting relationship between thermodynamics theory and architectural space form. And usually, even if there are some thermodynamic prototypes including the atrium, chimney, channel, etc., it is also often due to factors like functional streamlines rather than pure thermodynamics itself. On the other hand, the technical disciplines like HVAC are huge in their own right, and the application of architectural is only one of the branches of several major directions, and to promote the integration of disciplines, environmental engineers need to have a considerable understanding of architecture, and there should be a crossover between curriculum and teaching personnel. Insufficient understanding of the basic knowledge system of HVAC and other architectural techniques will lead to an insufficient theoretical basis for thermodynamic design in the education courses. So, Professor Ábalos in the first two years of the thermodynamics design course spent nearly half of the semester letting students explore and study the basic knowledge of thermodynamics, and he invited Mathias Schuler and other building technology field experts to teach the principle knowledge.

Secondly, it is the limitation of the form-oriented thermodynamics by itself. From the course designing process and the results of several years, it is not difficult to find many of the thermodynamic prototypes amplify some passive ecological strategies; including horizontal or vertical wind duct, narrow-below-wide self-shading retreat platform, and the building volume according to the cutting of sunlight and wind direction. This graphical form strategy is quite conceptual; it weakens function, structure, construction and many other practical factors. You can say it challenges the traditional paradigm way of thinking about architecture, and it is difficult to avoid contradiction between its form and the teaching objectives of conventional architecture.

2. The limitations In Practice

In general, an offshore design agency will invite eco-consultants to participate in the initial stage of a project for ecological design requirements. Well, domestic design project owners or architects seldom integrate ecological design in the conceptual phase, and the late intervention of the HVAC engineer is also relatively passive, just with the implementation of the basic HVAC equipment drawings, it is difficult to have a deep understanding of architectural space, let alone illuminating suggestions on form and space through thermodynamics in a positive manner at the beginning of the program. As a result, in practice, it is passive but fortunate to minimize the impact of equipment on space and form.

On the other hand, even if thermodynamics can provide a guideline design strategy at the beginning of a conceptual scheme, this ideal model is often dispelled in various ways by other demands of architecture and architectural practice. Different from teaching design, design practice cannot evade the demand and construction costs, the process of design is the process of repeatedly balancing the contradictions of all aspects. In general, the basic requirements should be met before the demands get satisfied. The passive enhancement of body comfort, which is advocated by thermodynamic buildings, is not a must in civil architecture. Even if some atriums or patios play a role in ecology, they are often designed to optimize space and facilitate evacuation.

### Improvement of Thermodynamic Design

Building thermodynamics is a part of the technical discipline of architecture, which is not a new subject. In the past five years, the Harvard School of Design has placed thermodynamics at the heart of the college syllabus, which shows that architectural design is trying to respond to the energy problems and challenges of today's social development and to arouse the attention of the education sector to the subject of architectural technology.

1. How should thermodynamic design retreat?

First of all, can we start with the idea popularization? From the vocational education to the architect, we may convey a more holistic concept of thermodynamics. When it comes to thermodynamic design, many people associate different concepts of green design, ecological design, overall design, sustainable design, or BREEAM, LEED, green building certification and other standards. Although they are often referred to and discussed, they are more like a lagging problem-solving approach and an additional list that must be completed, rather than a necessary clue in the design process.

In contrast, the combination of structure and architecture seems to have been more excavated and the building effect is more noticeable. Although thermodynamics itself is not as fundamental and significant as the role of the structural system in the construction of space, its influence is long-standing and ubiquitous, which affects people's feelings actively or passively. Should we also incorporate building technology into the basic teaching of architecture as closely as the structure and space construction? Is it possible to incorporate the form of exploration into the technical curriculum and incorporate the elements of technology into the design curriculum? If the concept of integration of technology and architecture can be implanted in the undergraduate course, it may reduce the differences of position and idea in practice of various types of work.

2. How Should Thermodynamic Design Be Returned?

Another question is how to find richness and meaning in addition to the rationality of thermodynamics. Continuing the analogy between thermodynamics and structure, if it is relatively easy to achieve structural expressionism, does the thermodynamic expressionism itself have similar rhythmic beauty? Or to further question, how much scope and depth should thermodynamics be represented? Should it be manifested or hidden? If it is manifested, will it be able to avoid becoming decorative or symbolic? Is the cost of manifesting thermodynamics sustainable from the overall view of the building? These are the specific issues that need to be discussed locally. The direct application of the ideal prototype seems to simplify these issues and is not necessarily the right proposal in reality.

Therefore, the prototype design method of

the thermodynamic building provides a huge menu of how to find a suitable position between the advance and retreat and making the appropriate choice. They are the test paper of new tools for the experience and judgment of the architect. The application of thermodynamics in practice needs to be returned to the thought about the nature of architecture, seeking a balance between rationality and meaning, or further transforming rationality into space meaning (which is the same as the relationship between structure and architecture). The purpose of thermodynamics as an enlightening tool is not to manifest the form itself, but to make it a type of existence that can be perceived and figured out by its deep integration. These reflections on the atmosphere, spatial quality, and physical experience are difficult to embody in teaching and ideal models but are particularly important in a built environment. How to create a beautiful physical atmosphere (temperature, wind, light), while reflecting on the meaning of immaterial, this may bring further direction and expansion to the design of thermodynamic architecture.

# 热力学建筑原型课程设计与思考

郑馨、郑思尧、吕欣欣

“热力学建筑原型设计——光”，是以上海图书馆浦东分馆设计竞赛为背景、具有真实城市环境与任务书要求的课程设计。近年来，热力学话题逐渐成为国内外建筑学界的热点，越来越多的学者致力于研究建筑形式与能量、物质之间的关系，反思大规模城市化、现代化对于气候、环境的影响。在科学技术不断发展的同时，建筑的建造与机械系统所消耗的有效能也比以往要多。本课题最终希望我们以能量的流动循环为出发点，以光作为关键词，提出新的建筑原型，响应气候和环境的变化。

课程设计分为四个部分：图书馆案例研究及基地调研、原型与环境模拟验证、原型植入场地，以及材料节点设计。首先，通过对全球图书馆案例的研究，并结合其回应气候的方式，对图书馆这一建筑类型有一个整体的认识。而后找到设计应对气候的关键词并得出热力学原型，再将对原型的组合或变形植入场地中得到建筑，最后需要对建筑的材料及节点进行详细设计。

## 图书馆案例研究及基地调研

与我们之前的设计课不太一样的是，李麟学和周渐佳两位老师非常重视前期的研究和思考，并不急于让我们直接进入设计环节。在正式上课之前，我们需要先选择一个图书馆案例，研究其处理光的方式以及应对气候的策略，从而找到相应的关键词，以便对于之后的设计有所启发。除此之外，初涉热力学领域，阅读也是非常重要的。因此最开始的一个月，我们就致力于案例研究和相关的学术阅读。我们三人的案例分别是位于墨西哥的巴斯孔塞洛斯图书馆、位于美国东北部的埃克塞特图书馆以及位于德国的柏林自由大学语言学图书馆。

巴斯孔塞洛斯图书馆从外到内，有不同的孔隙和层次来吸纳或者阻挡光；埃克塞特图书馆中也有从外到内根据不同功能的光线需求设计建筑引入光线的方式；柏林自由大学图书馆则是在半球体的空间中创造微气候，并通过建筑表皮材料的设置满足不同空间的光线需求。我们根据图纸重建数字模型，制作手工模型，并试着从气候、环境、功能等角度去解读光是如何运用于建筑中的。除了详细研究的三个案例，我们还把精选的世界各地图书馆案例根据纬度高低整理起来，并对它们的场地关系、体量、平面、立面、表皮（孔隙）等因素进行比较分析。我们发现，图书馆这一建筑类型跟同一地域其他建筑类型相比，需要有更适应当地气候的开孔方式，且总体而言，随着纬度的升高，孔隙率逐渐变小。一些使用了节能技术的建筑跟同纬度没有使用节能技术的建筑相比可以有更大的孔隙率。

图书馆案例研究为我们下一步原型的设计提供了想法。我们发现，孔隙所指的建筑表面的透光率、通道所指的光传输的路径，以及分层所指的建筑不同部位功能与环境的相互影响，可以作为我们图书馆设计的核心。因此，我们以孔隙、通道和分层作为关键词，进入下一步原型的设计。

## 原型与环境模拟验证

在案例研究的同时，我们还进行了基地调研。我们的基地位于浦东新区迎春路，现为浦东新区文化艺术指导中心。世纪大道的尽端，基地西北侧为浦东新区行政办公中心，东侧是浦东新区展览馆和银联大厦；南侧为公共绿地，隔锦绣路是世纪公园。场地给我们最直观的感觉就是城市密度极低，城市尺度巨大，而且绿树成荫。

在与老师几轮讨论后，我们觉得树木的能量机制可以作为我们研究光和能量的切入点。一棵树对阳光的利用率可以高达 95%，是现有的任何节能建筑都无法比拟的。于是，我们小组三人决定以树木的能量机制中孔隙和层次的因素为重点进行研究，“光合作用”这个题目也就此诞生。

我们一边研究植物的光合作用和呼吸作用的原理，一边开始阅读热力学相关的理论书籍，如基尔·莫的 *Convergence*、威廉·布雷厄姆的《热力学叙事》，对熵增、热库、循环等等概念有了一定的认识。我们希望将提炼概括出的植物能量流动循环的原理运用到建筑中去，创造一个可以模拟植物利用光能的过程的热力学机器。

我们将树对能量的利用划分为六个机制，分别是捕捉、转化、传导、传输、储存和输出。在树中可以找到不同的部位对应不同的机制：捕捉由树冠和树根共同完成，树冠吸收光能，树根则从土壤中吸取养分；能量从光能到化学能的转化在叶片中完成；树根汲取养分后自下而上传导到其他部位中；树干负责能量的运输；最后再由树冠对外界进行能量的输出。不同气候条件适宜不同的树木生长，同一种树木在不同的气候条件中某些部件的形态会有一定的变化来适应环境。经过分析，我们总结得出影响这些机制的因素。比如，表皮孔隙率、开孔排布方式和太阳能板、采光锥这样的设备可以捕捉光能，管道序列、口径和排布则可影响光在建筑中的传导。建筑中的一些因素也可以和植物对应起来。比如表皮孔隙率与树冠孔隙率相似，都起到捕捉光能和输出能量的作用。通过对植物性状的研究，我们可以找到适宜上海所处气候带的植物性状。同样地，也可以找到适宜不同方向、不同季节的建筑性状，作为建筑设计中需要遵守的基本准则。

我们的第一次尝试是把植物中负责六个能量机制的部件一一对应到建筑元素中。我们找到了对应功能的预制设备，比如负责能量捕捉的太阳炉、负责能量转化的热泵等，然后将它们组装在一起，就有了“热力学机器”第一稿。第一次汇报，我们的成果虽然在形式上很吸引眼球，但是经不起推敲。问题出在我们没有从原理和系统上研究能量的流动循环，只是进行简单的部件拼贴。我们开始明白，热力学设计的重点并不是做到与原理一一对应，而是要让能量机制在建筑中运作起来。

第一阶段汇报之后，我们还发现，现在的节能建筑及节能设备对于能量的利用是很有限而且不成系统的。于是，我们回到出发点，重新研究植物的特性，比如通过孔隙捕捉阳光，通过通道传递能量，在不同的层次上对能量有不同方式的运用。在老师的启发下，我们决定做一个上层有开孔、下层有风塔状通道，上下层交叠的原型。

我们通过改变参数，如孔隙的大小、通道的形态、不同层的厚度等，做出了一系列原型模型来研究这三个关键词对光的影响，并在同济大学的光学实验室中进行了冬至、夏至、春分三个节气每天的自然光照模拟。有一些因素的改变对于光线有非常明显且有趣的影响，比如上层的厚度、通道的方向及高度、通道的开口大小等。另有一些影响不明显的因素，比如开孔的形状、下层的厚度等则被我们从需要改变的参数中剔除了。基于模拟所得的结论，我们对不同参数的原型进行了组合并生成了我们的中期成果。中期评图我们获得很好的反馈，老师们都觉得这个方案很有潜力。然而由于当时时间紧迫，对于原型在建筑层面的组合还没有进行深入的思考。

令我们意外的是，我们将中期成果整理后，按照课程的原定计划参与上海市重大文化设施国际青年建筑师设计竞赛（上海图书馆组），不仅入围，还脱颖而出获得了二等奖，成为当时最年轻的获奖者。

## 原型植入场地

中期过后，我们进入原型的深化阶段。通过对建筑中所需要的不同层次的光、风、公共空间的逐步推导，我们深化了原型在建筑层面上的定义：烟囱状的通高空间连通着原型的底层，使得底层和风塔内部有良好的通风环境，也给了底层特殊的光照效果——风塔底部犹如一个个漫游式阅读空间的“锚固点”。上层的体量与风塔隔开一定的距离，拥有相对稳定的风环境和热环境，同时让光能够从体量的开孔处投射到中间层。而底层和上层之间的中间层则随着季节、时间的变化，形成供人漫步的风通道和多样的光影效果，并作为城市公园对公众完全开放。

我们确定了原型的五个变形因素，分别是风塔的高度、方向、形状，风塔与上层脱开的距离，以及上层的厚度。原型的不同尺度可以被赋予不同功能。XL 尺度可以是图书馆的塔楼，L 尺度是内部可通行的大的采光天井，M 尺度可以是阅览空间的采光井或展览空间的聚光井，S 尺度则是作为交通核，最小的 XS 尺度可以成为采光天窗。

基于原型研究中所确立的基本准则对不同尺度的原型进行组合，得到建筑的基本形态。再通过光环境和风环境的模拟进行优化，最终得到这样的建筑形态：烟囱状风塔的整体倾向北侧，西南立面呈弧形，防止顶部过热的同时使风塔南侧产生庭院，保证体量内部通风顺畅；尺度逐渐变小，不影响北部采光的同时保证风的顺畅通过，也最大限度在中间层引入自然风；烟囱形状由收

“光合作用”方案模型

“光合作用”方案渲染图

口逐渐变为喇叭口，为局部屋顶提供遮阳；中间层由薄渐厚，在提供良好的南部景观视野和日照的同时阻挡冬季西北风；地下层南部抬高半层，前开下沉式广场引入东南风；西南侧绿地利用植物高度和密度设风道，使东南风易于通过，同时阻挡冬季西北风。

## 材料设计研究

与此同时，我们进入一个新的话题，即研究表皮的材料模块的构造和设计。我们希望延续风塔和开孔结合的形式，将表面的开孔也分为内外两层，并赋予轻微的转动。我们设计了一种带孔的拼接材料，并将材料赋给风塔的东南和西南两个立面。材料以风车状排列组合，并固定在顺应立面形态的龙骨上。在外侧，因中部口径变小，可以实现风的加速进入。内侧，因中部到内部口径又变大，白色光滑的曲面对内形成开孔，增强光的漫射效果。根据不同的功能需求，我们可以改变材料的口径、旋转角度和进深。另外，在置入建筑体量时，我们将下部的开孔和扭转放大，上部则缩小，希望能通过这种变化，一方面控制不同功能区域的光环境，另一方面增强风塔的拔风效果。为了真正理解这种材料的构造，我们用硅胶做了磨具，制成 1:5 的混凝土模块，并且 3D 打印了 1:20 的构造模型。

最终，我们实现了六个机制在建筑中的运行。通过不同尺度的风塔和表皮材料，根据不同功能需求捕捉光能；光加热了风塔的南侧立面，促进通风，实现了光能—热能—风能的转化；风塔的方向和排布实现了光能的传导；底层和中间层引入东南风进入建筑内部实现能量的传输；庭院中的树和在建筑中活动的人因光而活跃、生长，作为存储光能的一种方式；充满活力的城市公园和内部空间则作为建筑对城市的回馈。

## 课程思考

回看当时的我们，从一开始对原理的一知半解，到后来越来越理解建筑的热力学，并做出这么多成果，确实让那一学期非常有意义。课程之后，我们对建筑的形式有了新的认知：建筑的原型可以通过对物理原理的研究抽象得出，这种或许看起来像“怪物”的形式不再只局限在传统美学或者空间的层面，而是具有物理上的性能以及其背后的形式逻辑的。最后，感谢李老师和周老师在这一学期中对我们的指导，在我们“误入歧途”时及时把我们拉回来，在我们迷茫时给予最诚挚的鼓励，在我们得奖时比我们还要高兴！

参考文献

[1] 伊纳吉 · 阿巴罗斯. 建筑热力学与美[M]. 同济大学出版社，2015.

[2] 李麟学. 知识 · 话语 · 范式——能量与热力学建筑的历史图景及当代前沿[J]. 时代建筑，2015(2).

[3] 威廉 · W. 布雷厄姆. 热力学叙事[J]. 张博远，译. 时代建筑, 2015(2).

[4] Nordhaus, William D. The Climate Casino: Risk, Uncertainty, and Economics for a Warming World[M]. New Haven: Yale University Press, 2014.

[5] Bejan, Adrian, JP Zane. Design in Nature: How the Constructal Law Governs Evolution in Biology, Physics, Technology, and Social Organization[M].New York: Doubleday, 2012.

[6] Moe, Kiel. Insulating Modernism - Isolated and Non-isolated Thermodynamics in Architecture[M]. Basel: Birkhäuser, 2014.

# Design and Thinking for the Course of Thermodynamic Architectural Prototype

ZHENG Xin,
ZHENG Siyao,
LYU Xinxin

## Introduction

Design of thermodynamic architectural prototype: The course design features light with actual urban environment task requirements under the background of the design competition for Pudong Branch of Shanghai Library. Thormodynamics has gradually become a hot topic across the Chinese architecture area in recent years. The research objects include the relationships between architectural form and energy and material, introspection about urbanization and influences of modernization on climate and environment. With the continuous development of science and technologies, the building of architectures and the corresponding mechanical system are consuming more available energies than ever before. This course is ultimately intended to propose a new architectural prototype that adapts to the climate and environment changes, taking light as the keyword, based on energy circulation.

The course is designed into four parts, i.e. library case study, site investigation, prototype, and environment simulative demonstration, field implantation and materialization design. First, an overall understanding of the architecture type of library is got by studying library cases across the globe combining the way of climate adaptation. Second, keywords of climate adaptation are found with thermodynamic prototypes. Third, the architecture is completed by combining the above prototypes and field implantation. Finally, the materials and nodes of the architecture are designed in details.

## Library Case Study and Site Investigation

Different from the previous design courses, this course pays much attention to early-stage study and thinking without rushing into the design part. Before the official beginning of class, a library was selected for each of us to learn the way of light treatment and the strategy of climate adaptation. The corresponding keywords were found to enlighten the later design. In addition, reading is of great importance for green hands in the thermodynamics field. We have been devoted to the case study and the relevant academic reading in the beginning month. Tho cases for us are Vasconcelos Library, Exeter Library, and the Philological Library, Free University of Berlin respectively.

For Vasconcelos Library, plenty of pores and layers are used to absorb or block light from outside to inside; for Exeter Library, light receiving methods are designed according to different functions from outside to inside; for the Philological Library, Free University of Berlin, microclimates are created in hemispheroidal space and various light demands are met by specially designing building surface materials. We rebuilt the digital models according to the drawings, fabricated craft models and explained how to utilize light in architectures from the climate, environment and function points of view. Besides the above three cases, libraries worldwide were selected and organized by us according to their latitudes. Such factors as site relationship, mass, plan, facade, and surface (pore) were subject to comparative analysis. It was found that, compared with other architecture types in the same region, a library requires opening more adaptable to the local climate conditions. Generally speaking, porosity gets smaller with the rising of latitude. Architectures with energy-conservation technologies may have a larger porosity than those without around the same latitude.

The above case study provided us with thoughts for the design of prototype. It was found that porosity, referring to the building surface transmittance, channels, referring to light transmission routes and layers, referring to mutual influences of different functional parts and environment may be considered as the core of library design. Therefore, we take porosity, channel, and layer as keywords for the design of prototype.

**Prototype and Environment Simulative Demonstration**

Site investigation is carried out along with case study. Our] site is located at the Cultural Guidance Center of Pudong New District, Yingchun Road, Pudong New District. The Administration Center of Pudong New District sits to the northwest of the site at the end of Century Avenue; the Exhibition Hall of Pudong New District and UnionPay Building to the east; a public green area to the south, Century Park across Jinxiu Road. This site gave us a natural feeling of extremely low urban density with a large green area.

After rounds of discussions with our tutors, we thought that the energy mechanism of trees may be a starting point of to study light and energy. The sunlight utilization rate of a tree can reach as high as 95%, unparalleled by any existing energy-conservation architectures. Therefore, we decided to focus on the factors of porosity and layer in the energy mechanism of the tree, where the subject of photosynthesis came up.

We started to read the relevant theoretic references including *Convergence* by Kiel Moe and *Thermodynamic Narratives* by William Braham while studying the principles of photosynthesis and respiration, where we learned entropy production, heat reservoir, circulation and other concepts. We hoped to summarize the principles of plant energy circulation, and apply these principles in architectural design to create a "thermodynamic machine" which can simulate the process of photosynthesis.

We divided the energy mechanism of the tree into six sections, i.e. capture, conversion, transfer, transmission, storage, and output. Different parts of tree correspond to different sections of mechanism: capture is conducted by the crown and root jointly, in particular, the crown captures light energy and the root captures nutrients in the soil; conversion is completed by the leaves; transfer of nutrients from the root to the other parts is conducted through the tree; the transmission of energy is carried out by the trunk; energy is output outside via the crown at last. Different trees are fit for different climate conditions. For the same type of trees, some parts may have certain changes in form to adapt to different climate conditions. The influencing factors of the above mechanism were concluded by us. E.g. porosity, pore layout, solar panel, and lighting cone can influence light energy capture while pipe sequence, diameter and layout can influence light conduction in the architecture. Some aspects in architecture can be compared with the plant. For example, surface porosity is similar to crown porosity for light energy capture and energy output. Through the analysis of plant characteristics, we can find plants fit for the climate conditions in Shanghai. Likewise, we could find architectural characteristics fit for different directions and seasons as a basic guideline in architectural design.

We first tried to match the six mechanism sections with architectural elements. We found corresponding prefabricated equipment such as solar furnace for energy capture and thermal pump for energy conversion. We assembled this equipment into the first thermodynamic prototype. Though attractive, it failed to withstand scrutiny. Because we didn't research the energy circulation from the angles of principle and system, instead, we created a simply assembled product. We started to understand that, the key to thermodynamic design is not to find a match in principle but to find an energy mechanism suitable for the architecture.

We also discovered in the first stage of a report that the existing energy-conservation buildings and equipment have limited and unsystematic use of energy. As a result, we went back to the start to study the energy utilization of plant in different layers with different methods. Inspired by our tutors, we decided to build a prototype with pores on the top layer and a wind-tower type channel on the bottom layer.

We fabricated a number of models with varied pore size, channel form, and layer thickness to study the influence of the three keywords on lighting. A whole-day natural lighting simulation test for the Winter Solstice, the Summer Solstice, and the Spring Equinox was carried out in the optical laboratory of Tongji University. The changes of certain factors, e.g. thickness of the top layer, direction and height of the channel, size of channel opening showed pronounced and interesting influence on lighting. We deleted some factors such as pore shape and thickness of the bottom layer, which have minor influences, among the parameters. We combined prototypes with different parameters based on simulation to generate our mid-stage result, which turned out to be accepted by our tutors as a potential plan. However, we didn't have much time to consider the combination in respect to architecture.

Unexpectedly, this mid-stage result won the second prize in the Shanghai International Design Contest of Major Cultural Facilities (library category), making us the youngest winners of the contest.

**Field Implantation**

We came to the deepening stage after the mid-stage. We further defined the prototype in architecture by gradual deduction of light, wind and public space for different layers: a chimney-like high space connects the bottom layer, creating an excellent ventilation environment and particular lighting result for the bottom layer and the internal wind tower. The wind tower bottom is like scattered "anchoring points" of a roaming reading space. The top mass is separated from the wind tower to have a relatively stable wind and thermal environment and to allow light to penetrate the pores and reach the middle layer. The middle layer forms an urban park open to the public with a wind channel for wandering and diversified lighting effects, changing with the season and time.

We finally determined five prototype parameters, i.e. height, direction and shape of a wind tower, distance from wind tower to top layer and thickness of the top layer. Different sizes of prototype represent different functions. Size XL can be used for a tower building, size L for asizeable light courtyard with internal passage function, size M for light well for reading space or condenser well for exhibition space, size S for transportation core, size XS for a skylight.

The basic architecture form was obtained by combining different sizes of prototypes based on the basic criterion established in the prototype study. This basic form was optimized by simulating the light and wind environments for a final form: the chimney-like wind tower leans towards north and forms an arc shape in its southwest facade to prevent top overheat and create a courtyard on the south, ensuring smooth internal ventilation; its size gets smaller gradually without influencing lighting of the north part, for leading natural wind to the middle layer to the maximum extent; its diameter gets larger from bottom to top to provide sunshade for part roof; its middle layer gets thicker gradually to block northwest wind while providing excellent scenic view and sunlight facing south; the south part of the basement is elevated by half a floor to introduce southeast wind; the green area in the southwest covered by high and intensive plants can introduce southeast wind and block northwest wind.

### Materialization Design

In the stage of material structure study and design, we expected to create two-layer surface poring as above with gentle rotation. We developed a pored splice material for the southeast and southwest facades of the wind tower. The material was arranged in a windmill shape and fastened to the keel adapted to the facade. The middle diameter gets smaller towards outside to let the wind in faster. The middle diameter gets larger towards inside to form an internal opening on the smooth white curve, enhancing light diffusion. We can change the diameter, rotating angle and depth of material according to different functional needs. Also, we enlarged the opening and rotating angle in the upper part and narrowed that in the lower part in order to control the light environment in various functional areas and enhance the updraft performance of the wind tower. We used silica gel mold to fabricate a concrete module in 1:5 and created a structural model in 1:20 by 3D printing.

Finally, we can run the six mechanism sections in architecture. Light is captured by the wind tower and surface material in different sizes based on various function needs; light-heat-wind conversion is realized as the south facade is heated by light and ventilation is increased; transfer of light energy is realized due to the direction and layout of the wind tower; internal energy transmission is realized by introducing southeast wind from the bottom and middle layers; trees in the courtyard and people inside the building grow and live as a method of storing light energy; an energetic urban park and internal space is so created as a gift of architecture to the city.

### Reflection

We have experienced a significant term from knowing little of thermodynamics to making great achievements by understanding and applying thermodynamic principles in a more skilled way. We have learned some new knowledge in respect to architecture form: an architecture prototype may be obtained by extracting physical principles. This "grotesque" prototype will not be limited by traditional aesthetics and space but be enriched by physical performance and logic. At last, we would like to thank Professor Li and Professor Zhou for their patient instructions and sincere encouragement to help us correct our mistakes and clear our doubts.

# 基于高密度环境的热力学城市原型研究

夏孔深、范雅婷

## 新加坡 2050 年的风险与挑战

本次竞赛选址于新加坡，要求以“Everyone Contributes”为主题在 1 平方公里的土地上设计容纳 10 万人居住的城市。其设定的城市建成时间为 2050 年，即设计必须对新加坡 2050 年将面对的风险与挑战进行预判。我们将新加坡未来的风险归纳为四个部分：Resource（资源）、Industry（产业）、Society（社会）、Kinetic（能动性）。

1. 资源

新加坡作为一个城市国家，各类资源极度缺乏，这一点对其未来的发展甚至国家安全都造成极大的威胁。在能源方面，新加坡现有能源结构以化石能源为主，99.6% 依赖进口；在淡水资源方面，大量依赖从马来西亚进口，一旦马来方面停止供应，整个新加坡将立即陷入生存危机；在土地资源方面，大面积填海造陆以及对土地的集约利用是目前新加坡应对土地资源紧缺的主要策略。

2. 产业

新加坡的高速发展很大程度上依赖其位于马六甲海峡的重要地理位置。发达的国际贸易带来了密切的商业文化信息交流，使得新加坡能够不断进行产业转型，从最初的商品贸易，到加工业，再到后来的金融服务业。而面向未来，新加坡也希望在信息技术、文化艺术、科技产业等各个尖端领域走在世界发展的前列。

3. 社会

在多元混合的社会结构下促进各民族和谐共处，建立统一的国家认同感，一直以来都是新加坡政府努力的方向。新加坡公共组屋政策以法律的形式规定每个组屋社区单元都要有各个民族的居民占比，华人占 74.3%，马来人占 13.3%，印度人占 9.1%，以保证不同民族的人生活在一起，并保持密切的联系。不仅仅是居住，在工作单位、学校等各处都有类似的政策建立共融的社会体系。

4. 能动性

正如我们前往新加坡考察时新加坡国立大学负责接待的教授所介绍，新加坡就像一辆自行车，必须处于不断的向前运动、不断的动态发展才能够保持稳定，一旦停下来便会倾覆。作为一个城市国家，其小规模的特征使其必须不断保持动态进步才能不被世界发展的洪流所淹没。而另一方面也正是因为其小，整个国家容易被迅速调动起来，迅速转型以适应瞬息万变的国际形势。因此以发展的眼光考虑城市设计，永远为未来的发展留有余地的设计才能够保证新加坡的能动性。

## 热力学原型研究

新加坡位于赤道附近，属于热带雨林气候。其高温、多雨、潮湿的热力学环境对当地的自然、传统建筑产生了很大的影响。我们向自然与传统学习，提取了自然原型热带雨林以及传统原型店屋，分别对其做热力学研究，并从中提炼指导热力学城市原型的基本思路。

1. 自然原型：热带雨林

热带雨林作为自然生态系统，其基本特征的形成充分体现了对于新加坡当地热带雨林气

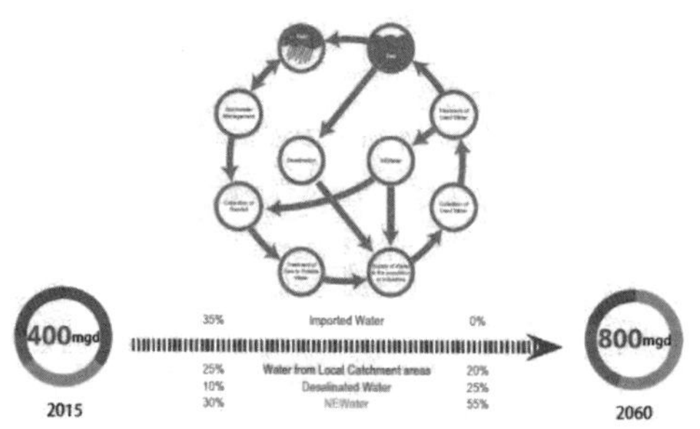


图 1 新加坡用水现状及规划

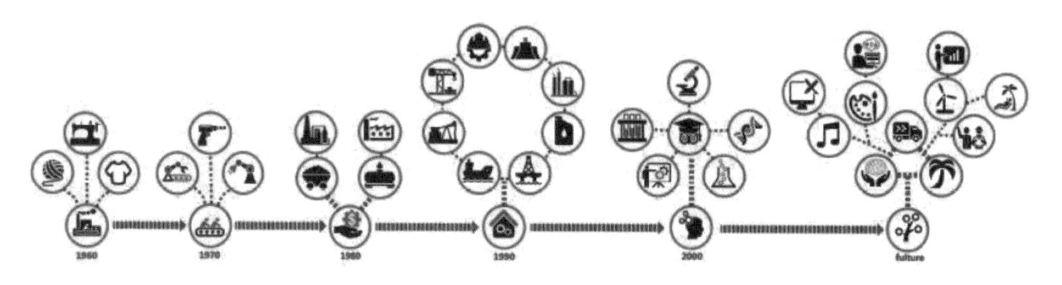

图 2 新加坡填海现状及规划

647
2015
?
766
2030



图 3 新加坡产业转型历史

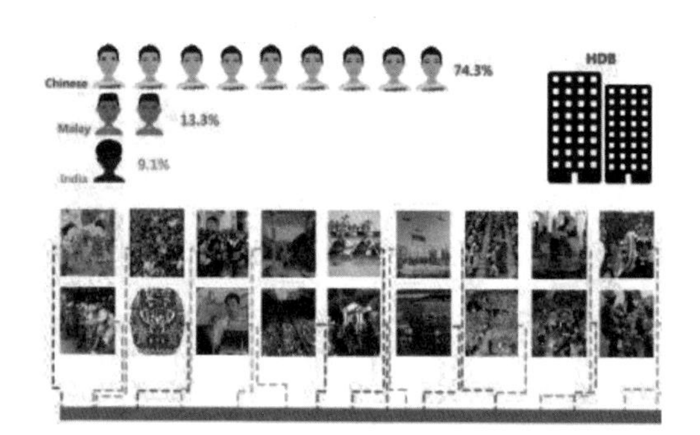


图 4 新加坡多元混合的社会结构

图 5 新加坡人口分布现状

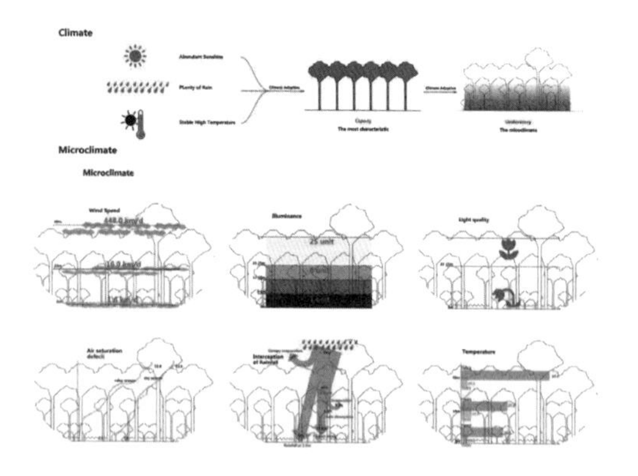


图 6 热力学自然原型：热带雨林

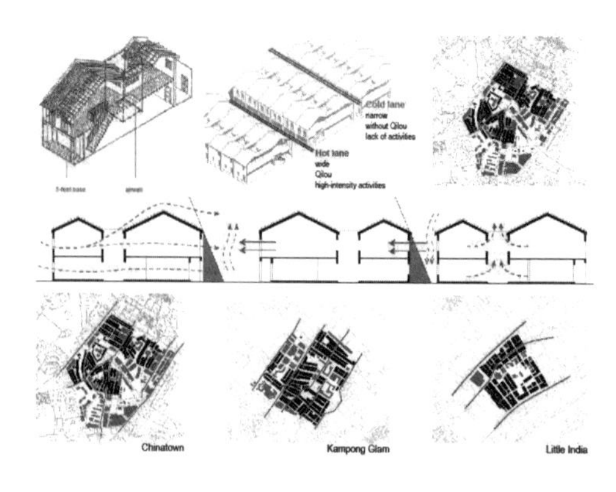


图 7 热力学传统原型：店屋

候的回应，呈现出自然的在地性。热带雨林的垂直结构从上而下依次分为冠层、林下叶层、矮乔木、灌木层、地衣苔藓层。树木为争夺更大的阳光照射面积拼命向上生长，在热带雨林的顶部形成浓密的冠层。冠层即是热带雨林最主要的太阳能捕获和能量生产区域，也为下部空间创造了相对稳定的微气候环境。冠层不但阻隔阳光，还阻隔了降雨、风速、高温，使得下部空间保持在一种阴影的微风环境中，且几乎不受外界气候变化的影响。

热带雨林丰富的垂直结构是与其分布式能量流动模式及相对均质的水平结构相对应的。由于热带雨林的主要能量来源为太阳能，冠层浓密地水平展开从而达到最大的阳光照射面。而其内部能量流动的方式即分布式能源系统非常高效，每一棵树都能形成能量捕获、能量引导、能量消耗的能量环。相比于如今城市广泛的产能区远离耗能区、跨区域调配能量的方式，这种方式减少了运输过程中的能量大量损耗。

2. 传统原型：店屋

店屋是一种在东南亚非常普遍的商住楼形式，从东北方的台湾地区经由福建再到西南方的马来西亚都出现过。但新加坡模式吸收了英国城市形态的观念并发展出适应热带气候条件的特点。莱佛士于 1822 年规定，房子要有走廊，以确保沿街为公众提供遮阴避雨的廊道。在建筑尺度上，传统店屋呈现沿街面 6 米左右的窄开间，20 多米的长进深，中间布置天井拔风以及补充自然光。在街区尺度上，一幢幢店屋联排形成街区，并形成两两正面相对的前街以及背面相对的后巷。前街街道尺度适宜，有灰空间廊道，是主要的商业活动、人活动的场所；而后巷相对更窄，仅作为后勤通道。后巷接受阳光辐射比前街少，故形成冷巷，冷热巷体系促进建筑内的热压通风。而在更大的城市尺度上，我们比较研究了牛车水、甘榜格南、小印度（分别是华人、马来人、印度人的传统聚居区）的店屋街区形态，发现其主要街道基本沿着城市主导风向并呈现流畅的风道，可见传统的店屋原型在城市尺度、街区尺度、建筑尺度上都有气候设计的考虑。

## 流动城市：新加坡热力学城市原型

新加坡作为发达的热带城市国家，其活跃的社会创造力和特有的国家能动性使其在城市建设的各个方面都已经成为世界的典范，被誉为花园城市。新加坡有能力在解决自身发展问题的同时为全球城市发展提供范式。在这样的背景下，我们提出“FLOW CITY——热力学城市原型”，创造一个不仅仅为新加坡，而且为全球城市提供范式的热力学城市原型。结合对于新加坡当地店屋以及热带雨林两大原型的研究，我们提出 FLOW CITY 的四个重要特征：基于自然的气候适应，组织空气通畅流动的气流城市；基于非中心化的自组织方式，引导能量高效流动的能流城市；基于多样性的群落关系，建构人与人之间亲密交流的人流城市；基于新陈代谢的城市演替，设计灵活可变永续发展的流动城市。

1. 区域规划

在区域规划层面，我们首先保留了基地中央大面积的森林，由于其顺应城市的主导风向，所以能够作为城市风道对区域的气候调节产生积极的作用。所保留森林之外的 6.78 平方公里的部分作为城市建设用地。

基地所处的 PAYA LABER 地区现为空军基地，内部与周边城市隔绝，缺乏道路连接。为促进未来区域间的联系，我们在基地内部引入两条主干道，沿东西向横跨森林，作为联系两部分地区的城市要道。在 PAYA LABER 基地内部我们也设置了一条 LRT 环线，选取几个未来的城市中心设立站点，并与基地北部 2030 年规划的地

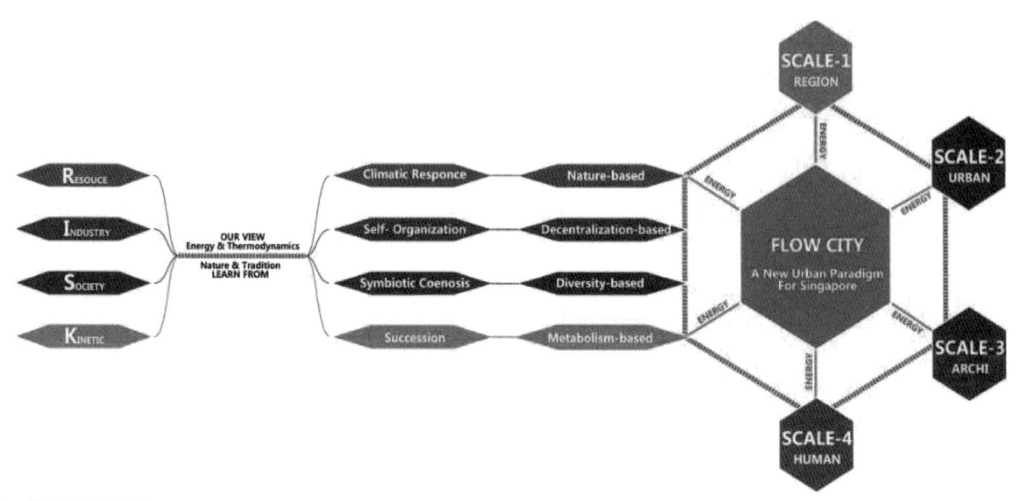


图 8　FLOW CITY 概念图

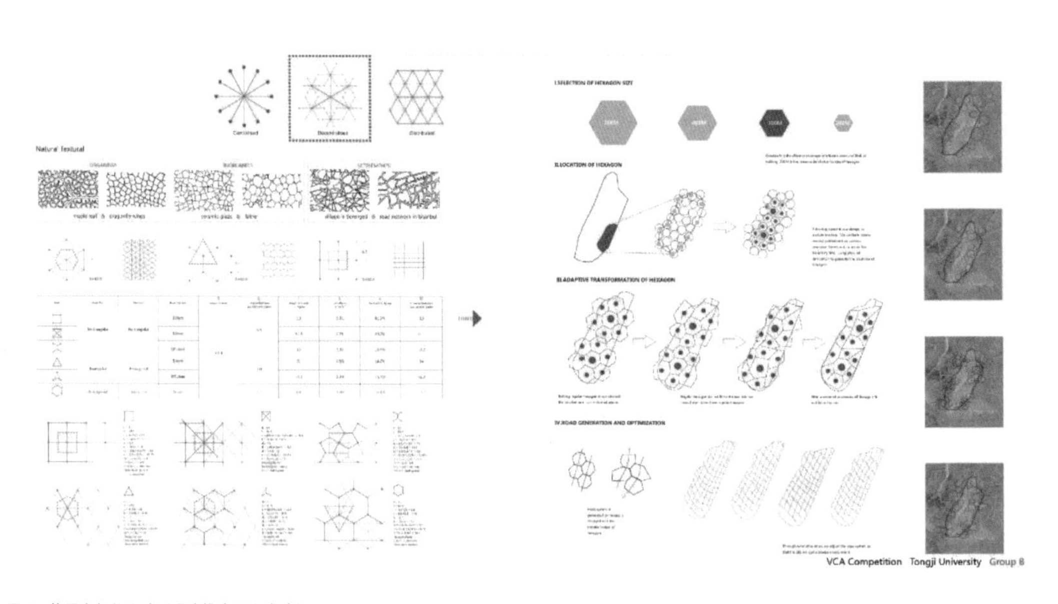


图 9　能量自组织六边形理论推演及生成过程

铁站以及基地南部正在修建的地铁线站点相连。

2．城市设计

在城市设计层面，我们借鉴热力学原型（热带雨林与马来传统建筑）的经验，首先顺应主导风向和次主导风向来组织菱形路网，形成城市风道并划定了城市街区形态。其次，我们学习热带雨林的群落结构来建立多元混合的社会关系，创造出一系列散落分布的高密度的混合功能群落。为了确定群落的尺度，我们参考了一些现存城市的街区尺度，如迪拜、曼哈顿、米尔顿 · 凯恩斯等，并且分析新加坡现存的街道空间状况，得出300 米的街区尺度为最优的组织模式。

我们学习弗雷 · 奥托在《占据与连接》中对于自然聚落自组织形态的研究，在自然界生物、非生物、人类聚落等结构中提取最常见的三种几何形——矩形、三角形、六边形，对其进行数学建模，根据占据与连接的方式不同，组合成六种基本的聚落形态关系。通过量化的数学计算比较各个参数，可以明确看出正六边形连接的网格具备能量传递的最高效性。基于此我们利用计算机软件建模，通过设定城市边界的排斥点，应用力学模拟形成人类聚落自组织布局的自然状态。通过能量六边形网格与风道菱形网格的相互契合，并根据基地边界条件做适当的修正，便得到了现有设计中的城市平面结构。而六边形体系作为能量流动的最高效解，成为城市水系统、城市智能电网系统、城市风塔系统、城市功能分区等各方面分布的基本原则。

3．建筑策略

在建筑策略层面，我们关注气候设计、原型结合、功能混合、自然渗透四个部分，总结出“匀”“混”“环”“透”“通”“漏”“皱”七点。

“匀”是建筑群体呈现均匀分布的状态，与热带雨林的平面均质化相类似，并且对应于热带雨林更为丰富的垂直结构，以及更为高效的分布式能量流动模式。

“混”是让单体建筑呈现功能混合的状态，包括建筑内部功能复合以及同一空间的混合使用。多元混合的空间在提高空间使用效率的同时也促进了空间气氛的营造，利于人与人之间互相交流，促进群落感与社区感，比单一功能的空间更有活力。

“环”是指物质循环，分别是屋面和垂直面的雨水收集系统、中水回用系统、太阳能利用系统、与绿植结合的碳循环系统。

“透”是指能量是在系统内部上下交换，达到更好的自组织的状态。新加坡本土的高层建筑通常也表现为通透、瘦小的体量，与环境进行更好的能量交换。

“通”是指建筑通风系统的设计，与城市尺度风道相对应有三种建筑通风形式，分别指捕风、辅助热压通风，以及捕风与热压通风结合。

“漏”是对雨林原型的回应，上述系统和雨林对太阳能和雨水的利用相仿。资源和能量在建筑中更合理地流动。

“皱”是冷热源的布置，建筑之间会形成很多窄缝，与较宽街区的充足阳光形成对比。小尺度的窄缝处适当布置植物，通过蒸腾作用增大这些部位的温差，来形成自然的热压通风。

## 结语

在我们设想的流动城市中，每个人、每座建筑、每棵树、每个物体都会参与能量流动的过程，形成能流、气流、人流、物流、信息流等。结合竞赛主题“Everyone Contributes”，每个个体都是能量的临时载体，在能量的捕获、引导、交换的过程中，有贡献更有收获。

我们尝试着基于自然的气候适应，组织空气通畅流动的气流城市；基于非中心化的自组织方式，引导能量高效流动的能流城市；基于多样性的群落关系，建构人与人之间亲密交流的人流城市；基于新陈代谢的城市演替，设计灵活可变永续发展的流动城市。

在这里，流动不仅仅是物质运动的方式，也是能量引导的途径，更是一种自组织的能动性。流动城市除了设计当下，也为未来城市的发展留出充足的余地，以便适应未来不可预知的各种变化并作出相应的调整，从而更好地迎接挑战，抵御风险，永续发展。

参考文献


[1] 李麟学 . 知识、话语、范式——能量与热力学建筑的历史图景及当代前沿 [J]. 时代建筑，2015(2)：10-16.

[2] 沙永杰 . 新加坡公共住宅的发展历程和设计理念 [J]. 时代建筑，2011(4).

[3] 布鲁诺 · 拉图尔 . 我们从未现代过 [M]. 刘鹏，安涅思，译 . 苏州大学出版社，2010.

[4] 丹尼尔 · 耶金 . 能源重塑世界 [M]. 朱玉犇，闫志敏 译 . 石油工业出版社，2012.

[5] 王才强，沙永杰，魏娟娟 . 新加坡的城市规划与发展 [J]. 亚洲城市，2012.

[6] 威廉 · W. 布雷厄姆 . 热力学叙事 [J]. 张博远，译 . 时代建筑，2015(2)：26-31.

[7] Will W.Braham. Architecture, style, and power: the work of civilization[C]// Will W. Braham and Daniel Willis. Architecture and Energy. Croydon, CRO 4YY: CPI Group(UK) Ltd, 2013: 9-24.

[8] Thomas Abel. Energy and the social hierarchy of households (and buildings)[C]//Will W.Braham and Daniel Willis. Architecture and Energy. Croydon,CRO 4YY: CPI Group(UK) Ltd, 2013: 49-63.

[9] John Thackara. Design in the light of dark energy[C]// Will W. Braham and Daniel Willis. Architecture and Energy. Croydon, CRO 4YY: CPI Group(UK) Ltd, 2013: 64-74.

[10] Shelton, B., Karakiewicz, J. and Kvan, T. The making of Hong Kong: From Vertical to Volumetric. New York: Routledge, 2011.

# A Study on the Urban Prototype of Thermo-dynamics Based on High-den-sity Envi-ronment

**XIA Kongshen, FAN Yating**

**Singapore's Risks and Challenges in the Year of 2050**

This competition in Singapore, themed "ev-eryone contributes", required to design a city of 100,000 people to live in an area of one square kilometer. The scheduled time for completion is the year of 2050, which means the design must predicate the risks and challenges that Singapore will face in 2050. We summarized the future "risk" of Singapore into four sections: R-resource, I-industry, S-society, K-kinetic.

1. R-Resource

As a city-state, Singapore's extreme lack of resources poses a great threat to its future development and even national security. In terms of energy, the existing energy structure in Singapore is mainly fossil energy, 99.6% of which depends on imports; freshwater resources largely depends on imports from Malaysia, if Malaysia stopped the supply, Singapore would immediately fall into a survival crisis; as for land resources, large-area reclamation and land use are the main strategies to deal with the shortage of land resources in Singapore at present.

2. I-Industry

Singapore's rapid development relies heavily on its close location to the Strait of Malacca. Advanced international trade brings close business culture and information exchange, which enables Singapore to continuously carry out industrial transformation, from the initial commodity trade and the processing industry to the later financial services. Facing the future, Singapore also hopes to be ahead of the world in various cutting-edge fields such as information technology, culture, arts, and science and technology industries.

3. S-Society

How to promote the harmonious coexistence of various nationalities and establish a unified sense of national identity under the pluralistic and mixed social structure of Singapore has always been the direction that the Singaporean government has been working on. In the Singapore public housing policy, it is stipulated in the law that every community unit of a group of flats must have a ratio of residents of all nationalities, the Chinese take up 74.3%, Malays account for 13.3%, Indians account for 9.1%. In this way, people of different nationalities can be ensured to live together and maintain close ties. Apart from residential houses, workplace, schools, and other places have similar policies to establish a harmonious social system.

4. K-Kinetic

A professor at the National University of Singapore told us that Singapore, like a bicycle, must be in constant motion, constantly moving forward to remain stable, otherwise it will capsize. As a city-state, its small-scale characteristics make it necessary to keep dynamic progress in order not to be overwhelmed by the torrent of world development. On the other hand, also thanks to its small scale, it is easy for the country to adapt to the rapidly changing international situation. Therefore, Singapore's initiative can only be ensured when the city is designed from the perspective of development with room for future development.

**Study on the Prototype of Thermodynamics**

Singapore is located near the equator and belongs to the tropical rainforest climate. Its high-temperature, rainy and humid thermodynamic environment has a great impact on the local natural and traditional buildings. We extract the natural prototype rainforest and traditional prototype store by learning from nature and tradition, thermodynamic researches were conducted respectively, and we concluded the basic idea of guiding the city prototype of thermodynamics.

1. Natural Prototype: Tropical Rainforest

Tropical rainforest as a natural ecosystem, the formation of its basic characteristics fully reflects the local tropical rainforest climate, showing a natural localism. The vertical structure from above down of tropical rainforest is divided into canopy layer, undergrowth leaf layer, dwarf tree, shrub layer, and lichen moss layer. Trees struggle to stretch out for more sun rays, forming a dense canopy at the top of the rainforest. As the main solar capture and energy producing part, the canopy has also created a relatively stable micro-climate environment for the lower space. The canopy not only blocks the sun and the rain, reduced wind speed and high temperature, so that the lower space maintains a shaded breeze environment, which is almost unaffected by the external climate change.

The rich vertical structure of rainforest corresponds to its distributed energy flow pattern and relatively homogeneous horizontal structure. Because the main energy source of the rainforest is solar energy, the canopy is densely spread horizontally to reach the maximum sunlight exposure. And how the energy is flowing makes the distributed energy system very efficient; each tree can form an energy ring of energy capture, energy guide, energy consumption. Compared to the vast energy-producing areas of human cities now far away from energy-consuming areas, the way to deploy energy across regions reduces energy loss in the transportation process.

2. Traditional Prototype: Shop-house

Shop-house is a prevalent form of commercial and residential buildings in Southeast Asia, from the northeastern region of Taiwan via Fujian to the southwest of Malaysia. But the Singapore model absorbs the concept of the British Empire and develops features adapted to tropical conditions (Lim, 1993). Raffles in 1822 stipulated that houses should provide sheltered corridors along the streets for the public. On the architectural scale, the traditional shop-house along the street presents a 6-meter-narrow bay, chin deep of 20 meters and above, the patio should be built in the middle to draw wind and replenish natural light. On the block scale, shop-houses are lined up face-to-face to form a front street, or back-to-back to form an alley. The front street with gray corridor has a suitable scale for commercial and human activities, the narrower back alley only serves as the logistical passageway. Because the back alley receives less sun radiation than the front street, the cold alley can be formed, and hot and cold alley system promotes the thermal pressure ventilation in the building. On the larger urban scale, we have studied the Chinatown of Singapore, Kampong Glam and Little India, which are the forms of the shop-house block in the traditional populated areas of Chinese, Malay, and Indian, and found that the main street is basically arranged along the dominant wind direction of the city, presenting a smooth air duct. It's clear that the traditional shop-house prototype has climate design considerations in urban scale, block scale and architectural scale.

**Flow City: Singapore Thermodynamic City Prototype**

As a developed tropical city-state known as the Garden City, Singapore's active social creativity and unique national initiative have site a model for the world in all aspects of urban constructio. Singapore can provide a paradigm for global urban development while addressing its development problems. In such a context, we propose "Flow City-thermodynamic city prototype" to create a prototype of a thermodynamic city that is not just for Singapore but for the global cities. Combined with the study of two major prototypes of Singapore's local shop-house and rainforest, we come up with 4 important characteristics of Flow City: Based on natural climate adaptation, organize air unobstructed airflow city; Based on the non-central self-organizing mode, guide high-efficiency energy flow city; Based on the diversity of community relations, construct communication-close people flow city; Based on the metabolism of urban succession, design flexible, variable and sustainable mobile city.

1. Regional Planning

At the regional planning level, we have reserved a large area of forest in the center of the base because it conforms to the dominant wind direction of the city and plays a positive role in the climate regulation of the region as the urban air duct. The other 6.78 Sq. kilometer is left as urban construction land.

In the PAYA LABER area, where the base is located, is now used as an air force base, internal and surrounding cities appear isolated, lacking road connections. To facilitate future regional linkages, we have introduced two main roads along the east-west axis across the forest to serve as a link between the two parts of the city. In PAYA LABER base, we've also set up an LRT link, select a few future urban centers to set up sites, and it is connected with the planned subway station in the north of the base in 2030 and the subway station under construction in the south of the base.

2. Urban Design

At the design level, we use the experience of the thermodynamic prototype (tropical rainforest and traditional Malay architecture) to organize rhombic road network and to form urban air duct and define city block form by conforming to the dominant wind direction and secondary wind direction. Secondly, we study the community structure of the rainforest to establish a pluralistic and mixed social relationship, creating a series of scattered and dense diverse functional communities. In order to determine the scale of the community, we refer to some of the existing cities, such as Dubai, Manhattan, and Milton Keynes and so on, and analyze the existing street space in Singapore and find out 300m block scale is the best organization model.

We study Frei Otto's study on the self-organizing form of natural communities in "Occupying and Connecting", three most common geometric forms are extracted from structures such as natural biological, non-biological, human communities: rectangular, triangular and hexagon. Mathematical modoling is conducted to find 6 kinds of the basic relationship of community form according to the different ways of occupying and connecting. It can be clearly seen that the grid connected by the regular hexagon has the highest efficiency of energy transfer through quantitative mathematical calculation and parameter comparison. Based on this, we use computer software to set models by setting the exclusion point of the city boundary, applying mechanics simulation to form the natural state of the human community's self-organization layout. The urban plane structure of the existing design is obtained by the combination of the energy hexagonal grid and the wind tunnel rhombus grid, and the appropriate correction according to the boundary conditions of the base. As the most efficient solution of energy flow, the hexagonal system has become the basic principle of the urban water system, urban intelligent grid system, urban wind tower system, and urban functional zoning.

3. Building Strategy

At the architectural strategy level, by paying attention to climate design, prototype combination, functional mixing, and natural infiltration, we summarized the characteristics of architectural strategy, as follows: "uniform", "mixed", "ring", "thorough", "pass", "leakage", and "wrinkle".

"Uniform" is a homogeneous distribution of the building population, which is similar to the surface homogenization of the rainforest, and corresponds to a more abundant vertical structure of the rainforest and a more efficient distributed energy flow model.

"Mixed" refers to the state of functions, not only of the building's internal functional compositions but also of the spaces. Mixed-used spaces are more energetic in the way that they enhance the efficiency, help build up space atmosphere, which further more will encourage communications between people and establish community sense.

"Ring" refers to the material circulation of the roof and vertical surface of the rainwater harvesting system, the water reuse system, solar energy utilization system, and green plant combined carbon system.

"Thorough" means how energy is exchanged within the system to achieve a better self-organizing state. Singapore's high-rise buildings are characterized by a transparent, thin volume, which has a better energy exchange with the environment.

"Pass" refers to the design of the building ventilation system, and the city-scale wind tunnel corresponding to the three kinds of ventilation forms respectively refers to trapping, auxiliary hot-pressing ventilation and the combination of wind-trapping and hot-pressed ventilation.

"Leakage" is a response to the rainforest prototype, which is similar to the use of solar and rainwater in the rainforest. Resources and energy in buildings can flow more reasonably.

"Wrinkle" is the heat and cold source of the layout; the building will form a lot of narrow seam in contrast to the abundant sunshine in the wider neighborhood. The plant can be suitably placed in the small-scale narrow seam with the purpose of an increasing temperature difference by transpiration effect, to form the natural hot pressing ventilation.

**Summary**

In the Flow City we conceive every person, every building, every tree, every object will participate in the process of energy flow, forming an energy flow, airflow, people flow, matter flow, and information flow. Combining the theme of competition "Everyone contributes ," each individual is a temporary carrier of energy, and in the process of capturing, directing and exchanging energy, it contributes more and gains more.

We attempt to organize air flow cities based

on natural climate adaptation; direct the energy-efficient flow city based on the decentralized self-organizing mode; construct the people's intimate communication based on the diversity of the community relationship; design flexible variable sustainable development of the mobile city based on the metabolism of urban succession.

Here, flow is not only the way of physical movement, but also the way of energy guidance, and a self-organizing activity. Flow City is not only a design for the present but also leaves plenty of room for future urban development, adapting to the unpredictable changes in the future, so as to better meet the challenge, resist the risk and achieve sustainable development.

近年来，热力学研究逐渐成为建筑学界关注的热点。通过在能量、物质与建筑形式之间的往复思考，热力学建筑试图重新建立建筑的当代设计方法与评价体系，以此反思大规模现代化之后对气候、环境这些因素考虑的缺失。这样的命题对于当代中国城市与建筑有着切实的意义。面对这样一个前沿话题，本系列课题采用哈佛大学设计研究生院的教学模式，通过专题研究、原型建立、软件模拟等抽象与具象的训练方式，对中国气候环境下的热力学原型加以探索。

热力学与很多概念有关，既有体积、体量这样随系统大小发生变化的外部属性，又有温度、压强、热容这样不随之变化的内部属性，这些都是本系列课题的讨论对象。而我们的环境中有很多因素都会影响建筑形式与对此的热力学评估，比如空气、通风、温度、材料等。2014 至 2018 年的五个学期，分别以“垂直城市”“空气”“光”“热”和“自然系统”作为关键词，各展开为期一学期的课程设计。

Thermodynamics has gradually become the focus of attention in the architectural field. Based on the reciprocal thinking of energy, material and architectural form, thermodynamic architecture attempts to re-establish the contemporary design method and evaluation system to compensate for the lack of consideration of the factors of climate and environment. Such a proposition is of great significance to contemporary Chinese cities and architecture. Faced with such a cutting-edge topic, it draws lessons of design training methods from Harvard University GSD. Through the abstract and concrete training methods such as special research, prototype study and software simulation, it explores the thermodynamic prototype in the climate environment of China.

Thermodynamics involves many concepts, not only the external attributes that vary with the system size, such as volume, size, but also the internal attributes which do not change with the system, such as internal temperature, pressure, heat properties. These are the topics of discussion in this teaching program. There are a number of factors in our environment that can affect the form of the building and the thermodynamic assessment of it, such as air, ventilation, temperature, materials. In the five semesters from 2014-2018, combined with the actual conditions of the base, on the basis of the establishment of a thermodynamic prototype, we have designed the curriculum for each semester around the following key words: Vertical City, Air, Light, Heat and Natural system.

# 设计
# 成果

# WORKS

# 运动中的空气
# Air in Motion

基地位于上海市西南角，曾经是虹桥经济开发区的地标性建筑，两侧分别是内环高架和延安路高架。现状中，虹桥宾馆与银河宾馆正在经历功能与业态的转换，两栋楼中间夹存的设备用房通过设备优化有待拆除，以达到用地的整合。本次设计的用地正是利用了这块要拆除的设备用房，两栋高层的存在对用地的物理环境有关键性的影响。

The site is located at the southwest corner of Shanghai, which had once been the landmark building of the Hongqiao Economic Development Zone, with both Inner Ring Road and Yan'an Road Viaduct in the current situation, Rainbow Hotel and Galaxy Hotel are experiencing the conversion of functions and formats. The equipment room in the middle of the two buildings remains to be dismantled. The former physical space is about to be integrated through the optimization of the equipment to the land. The purpose of the design is to make use of the equipment room which is going to be demolished. The existence of two high-rise buildings plays a key role in the physical environment of the land.

指导老师：李麟学、周渐佳
评审老师：张斌、高军、吴迪、胡琛琛、李卓
助教：李骜
学生：张润泽、周姝、李慧妮、潘亦欣、吴剑翎、丁一
Tutors: LI Linxue, ZHOU Jianjia
Review Teachers: ZHANG Bin, GAO Jun, WU Di, HU Chenchen, LI Zhuo
Teaching Assistant: LI Ao
Students: ZHANG Runze, ZHOU Shu, LI Huini, PAN Yixin,
Kimling Irene Ng Zheng, DING Yi

从左到右：塔风之下、风之甬道、浮游之森、风来风往模型照片
From Left to Right: Model photos of Under the Wind, Wind Path, Gardens on Gardens, In and Out

设计从对于中东地区风塔的科学性研究入手，在不同入风角度、风塔剖面形式、风塔数量组合等方面构建基础原型，热力学原型具有高度适应和形态控制的潜力。结合上海气候参数与基地气流现状，本案提出了一个风塔的“群组”，作为新的植入要素，最大程度改善场地的热力学环境，并提供一个富有魅力的能量形式化的载体。

The design starts with the scientific research on the wind towers in the Middle East and builds the basic prototypes in different aspects of wind direction, wind tower profile and wind tower number combination. Thermodynamic prototypes have the potential for high adaptation and morphological control. Combining the climate parameters of Shanghai with the current status of the base airflow, a "group" of wind towers is proposed as a new implant element to maximize the thermodynamic environment of the site and provide a fascinating carrier for energy formation.

# 塔风之下
# Under the Wind

学生 Student：
张润泽 ZHANG Runze

指导老师 Tutors：
李麟学 LI Linxue
周渐佳 ZHOU Jianjia

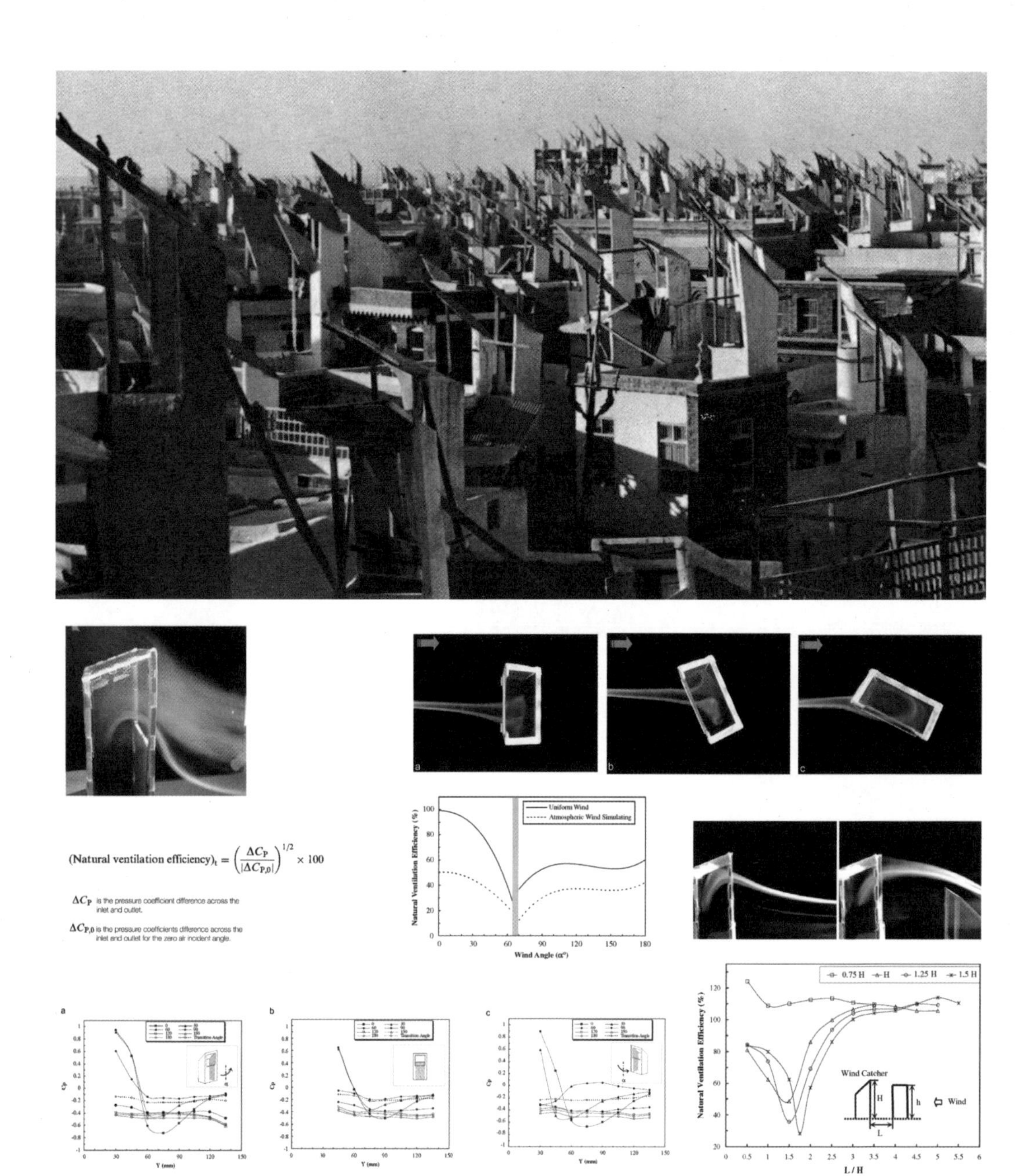


案例研究：中东地区风塔的定量关系
Case Study: Wind Tower in the Middle East

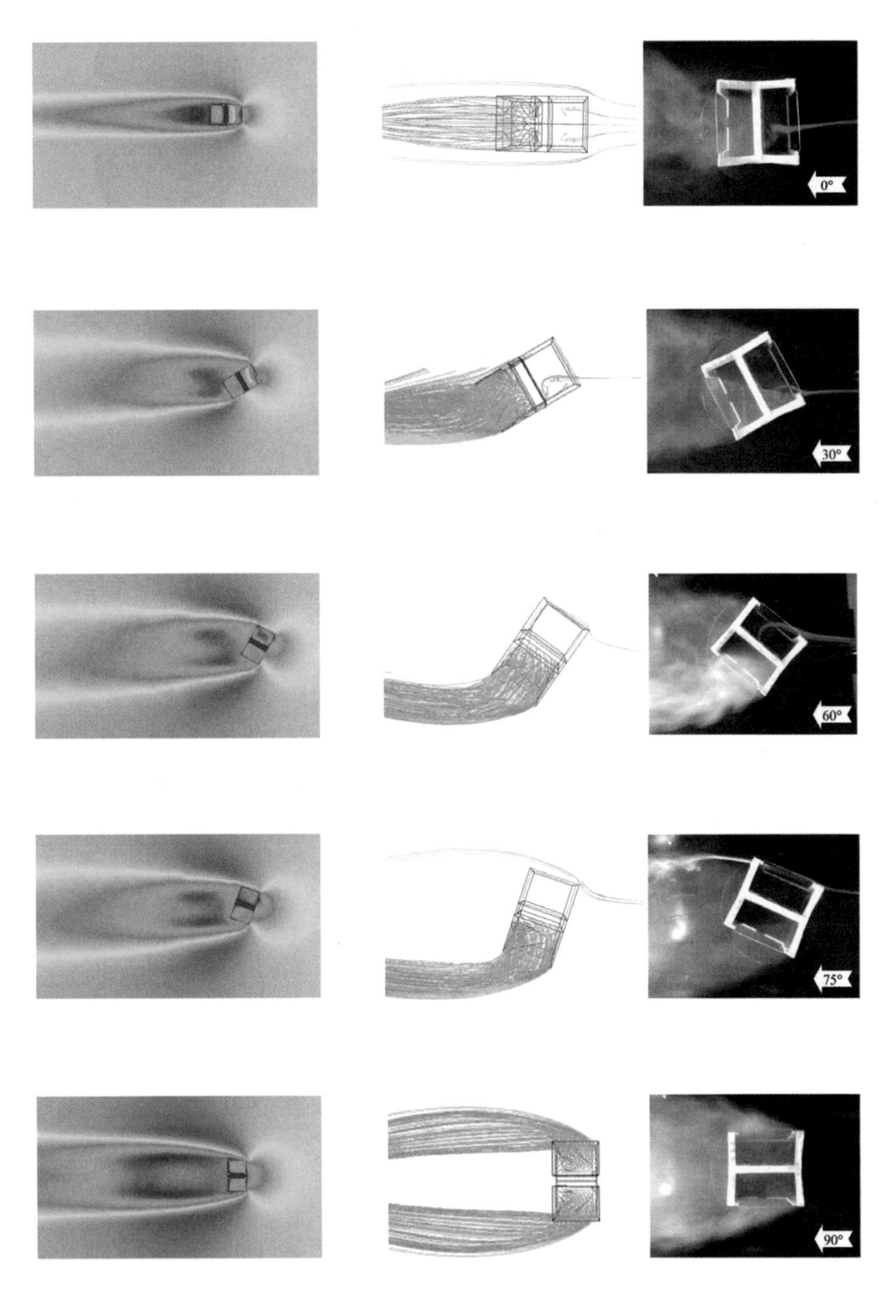


不同入风角度的风塔风环境模拟
Wind Environment Simulations of Wind Towers with Different Angles of Entry

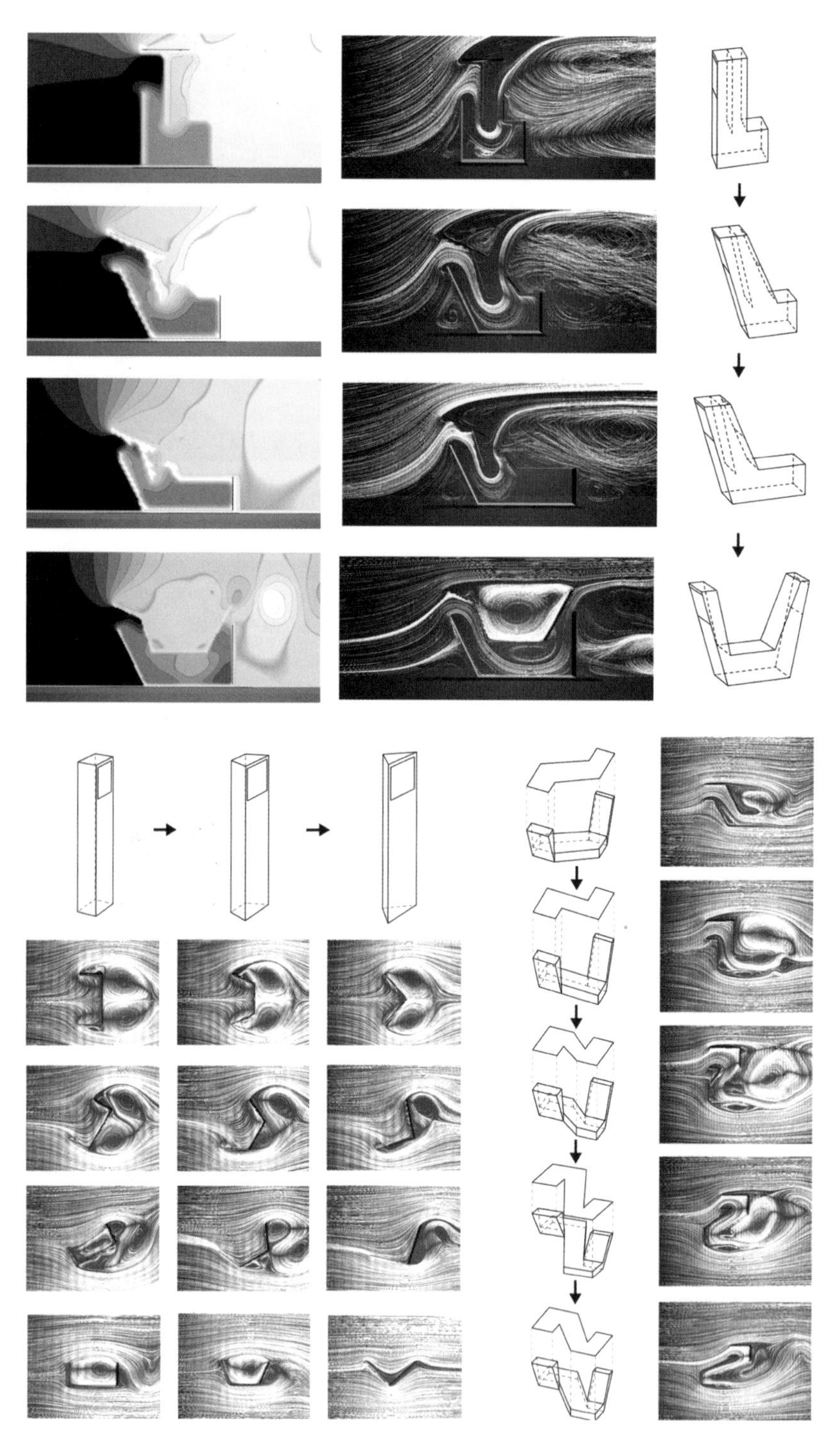

原型演变：热力学抽象

Prototype Transformation: Abstraction according to Thermodynamics

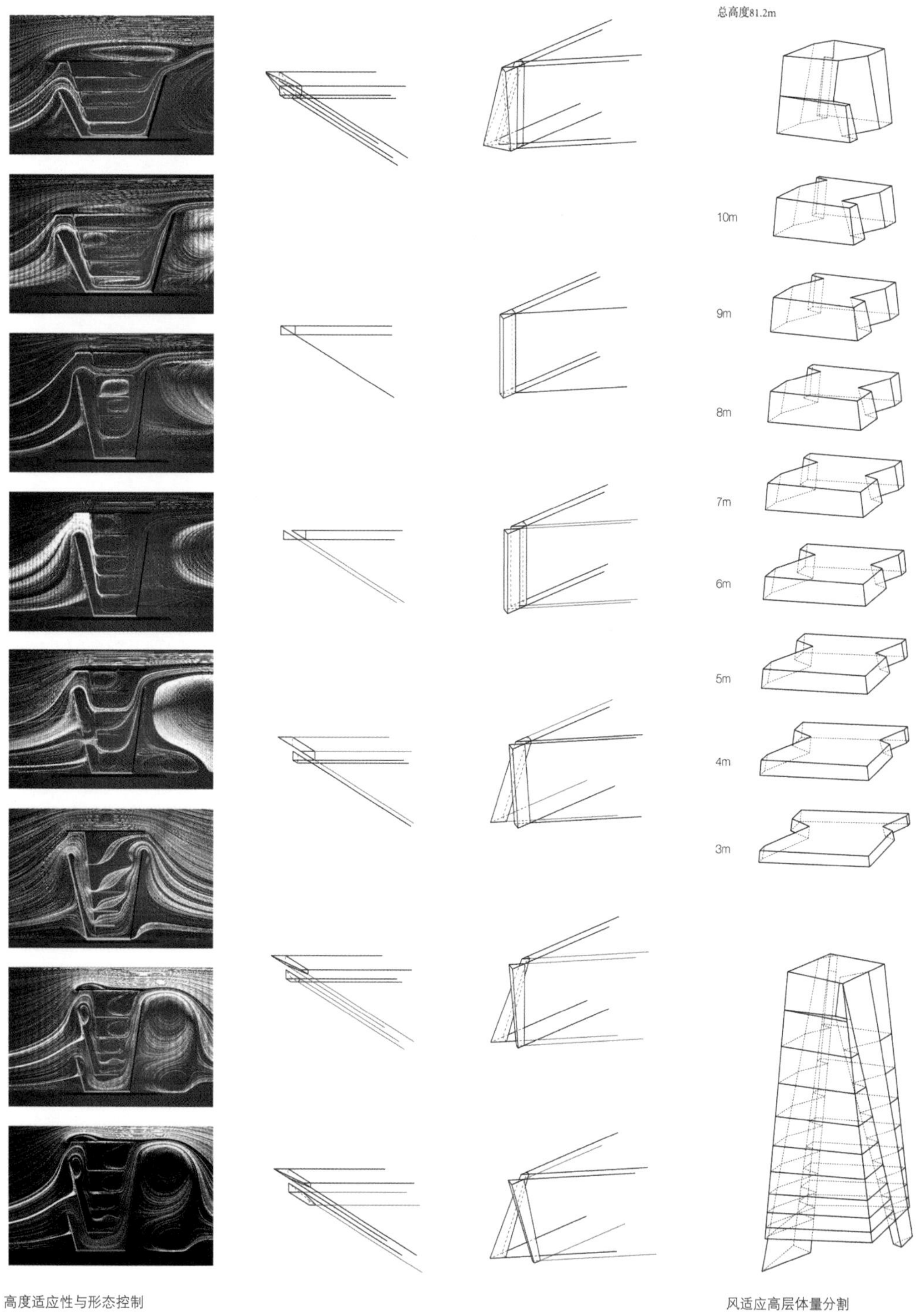


高度适应性与形态控制
Height Adaption and Morphological Control

风适应高层体量分割
Volume Division by Wind Flow

## 热力学建筑原型 Thermodynamic Architectural Prototype

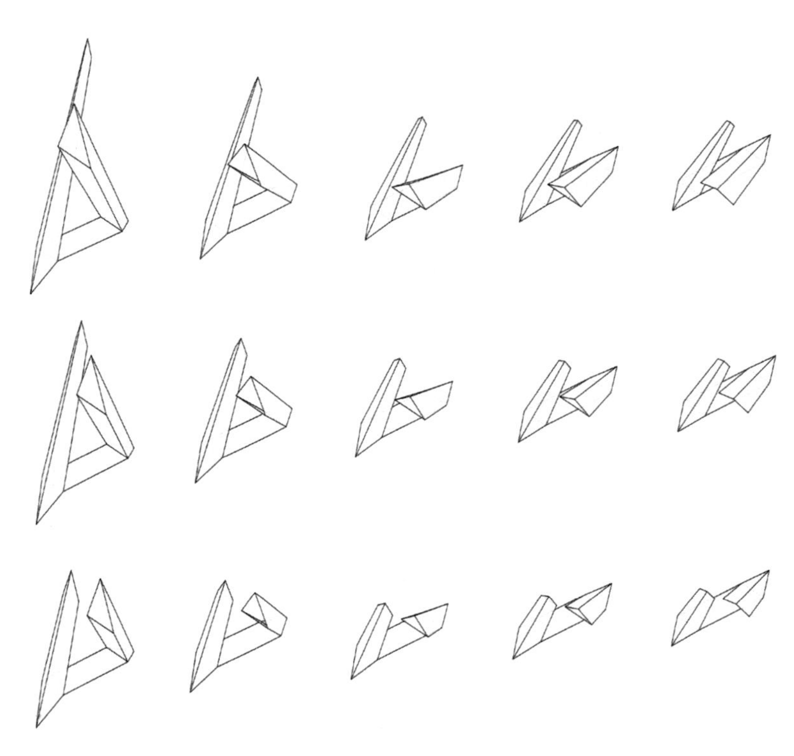

原型演变：包络曲线研究
Prototype Transformation - 3D Facade for Counter Test

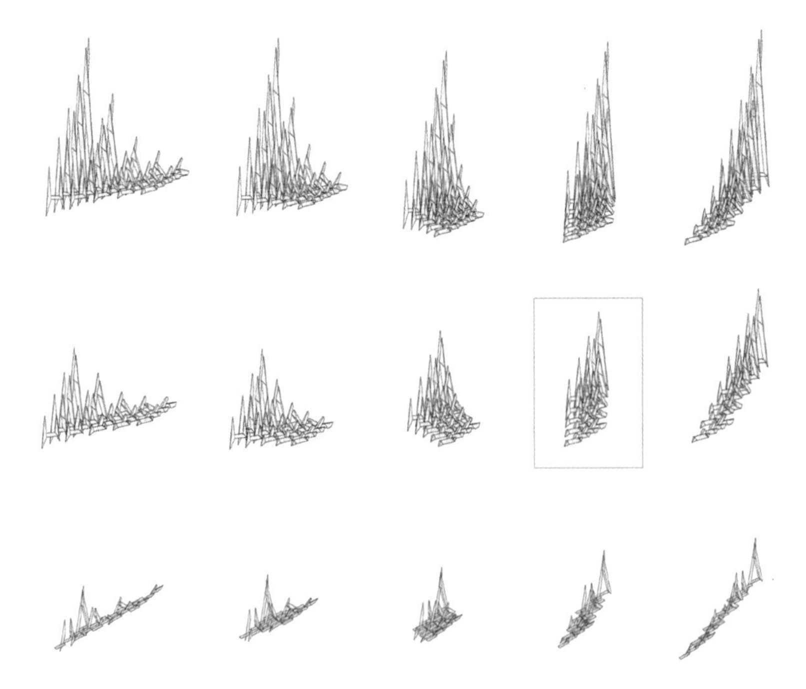

原型演变：三维肌理单体
Prototype Transformation - 3D Facade Units

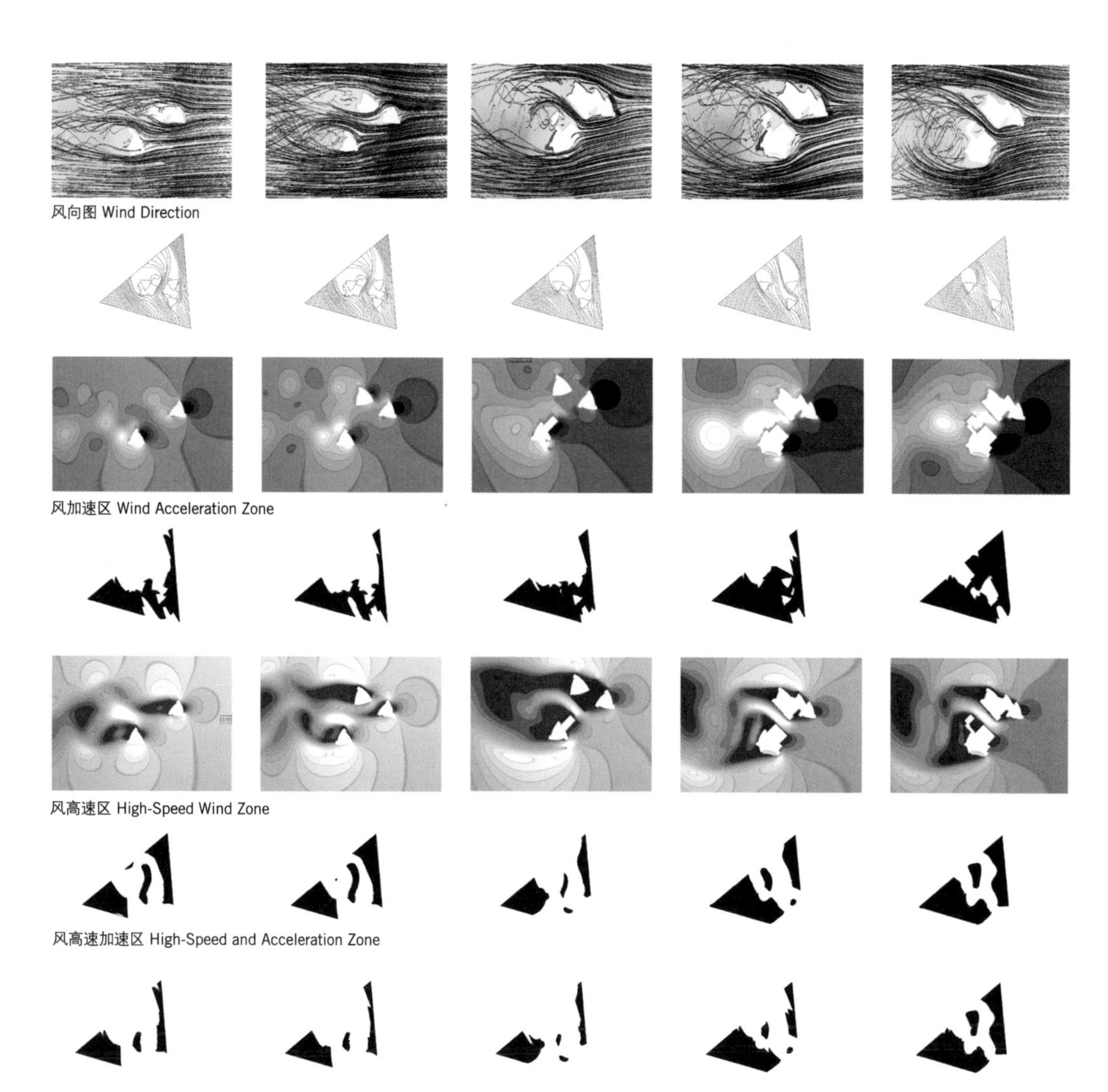


基地风环境分析：风塔适应区研究
Research of Suitability for Wind Tower in Site

## 热力学建筑原型 Thermodynamic Architectural Prototype

原型生成
Prototype Generation

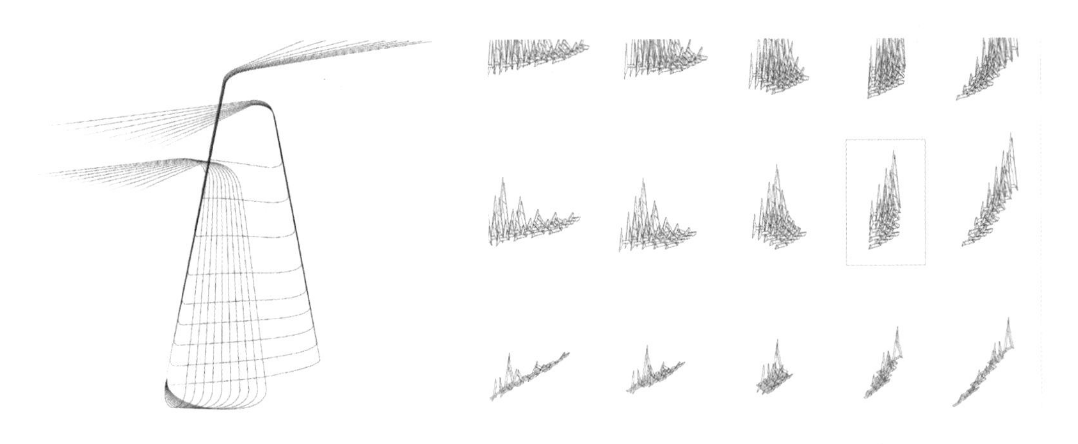

原型植入场地
Prototype Implantation

通过模拟高层塔楼之间风环境，呈现基地现有的微气候，并作为构建一个风的甬道的依据。多孔性与孔隙率成为构建甬道的控制参数，促成一个具有潜力的建筑形式。

Through the simulation of the wind environment between the high-rise towers, the existing microclimate of the base is presented and used as the basis for constructing a wind path. Porosity is the decisive parameter for constructing the path, contributing to a building form with potential.

# 风之甬道
# Wind Path

学生 Student：
吴剑翎 Kimling Irene Ng Zheng

指导老师 Tutors：
李麟学 LI Linxue
周渐佳 ZHOU Jianjia

案例研究：2000 年德国汉诺威世博会委内瑞拉亭
Case Study: Venezuela Pavilion, World Expo, Hannover, Germany, 2000

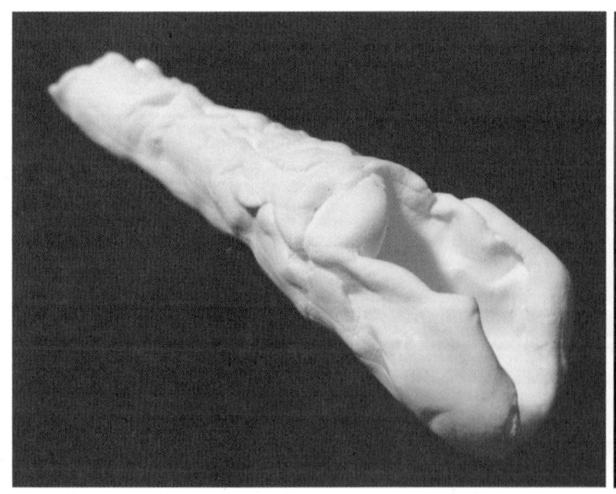

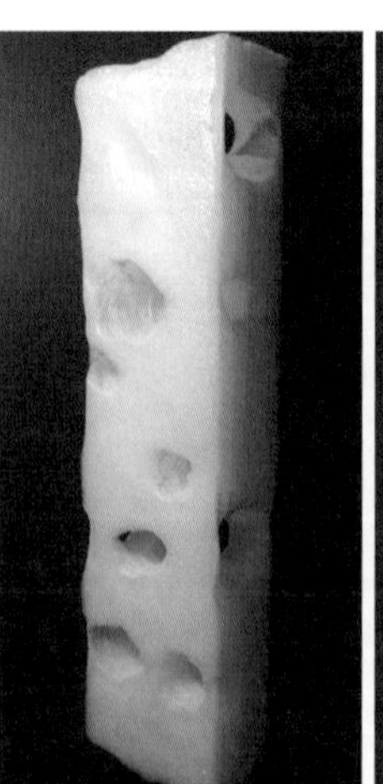

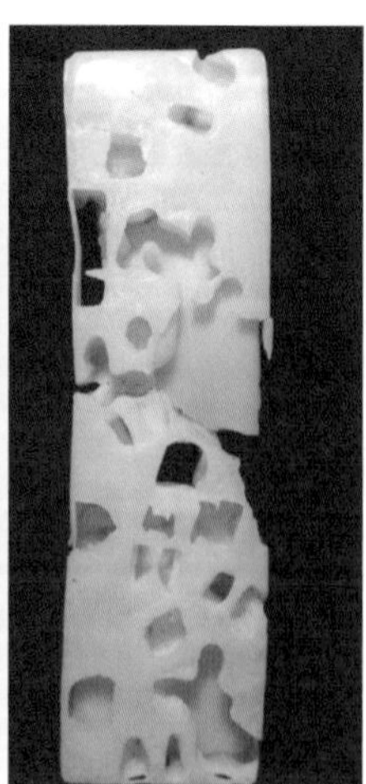

孔洞模型
Porosity Model

根据上海风环境，设计原型为东南—西北方向的圆柱形空心管。东南方向开口大，用以进风；西北方向开口小，减少通风量，由此引出“孔洞”的概念。通过开启不一样的孔洞，再运用软件模拟出风的轨迹，可以知道在建筑体量上孔洞的开启方向、孔洞大小和孔洞数量对建筑内部及外部风环境的影响。

According to the wind environment in Shanghai, a southeast-northwest cylindrical hollow tube is designed as prototype. The opening in the southeast direction is large enough for air intake; the opening in the north-west direction is small for reducing the ventilation volume. By opening different holes and using software, it is possible to simulate the trajectory of the wind and know how the opening direction, the size and the number of holes influence both internal and external wind environment on the building volume.

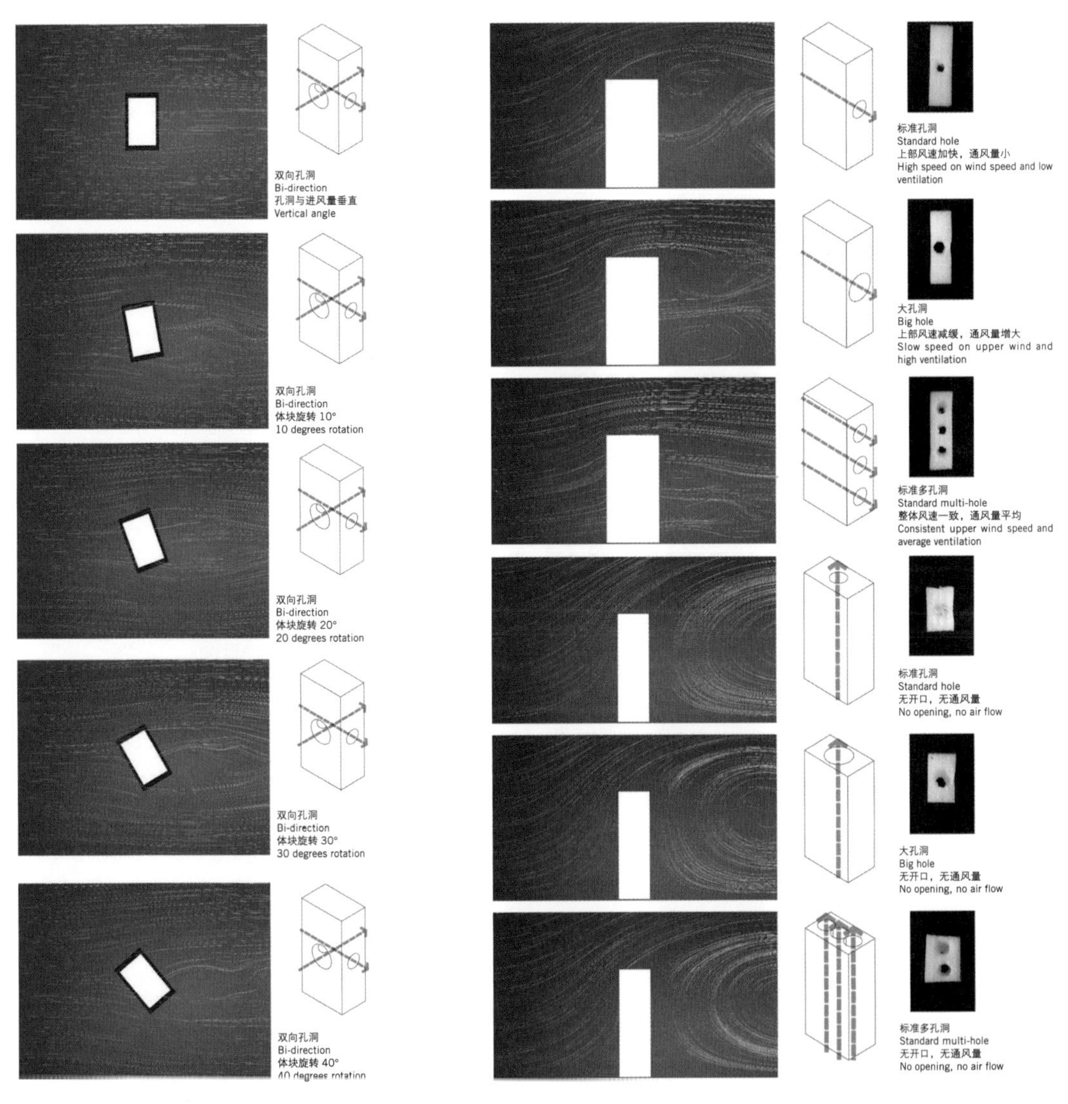


孔洞角度与风环境模拟
The Angles of Holes and Wind Environment Simulations

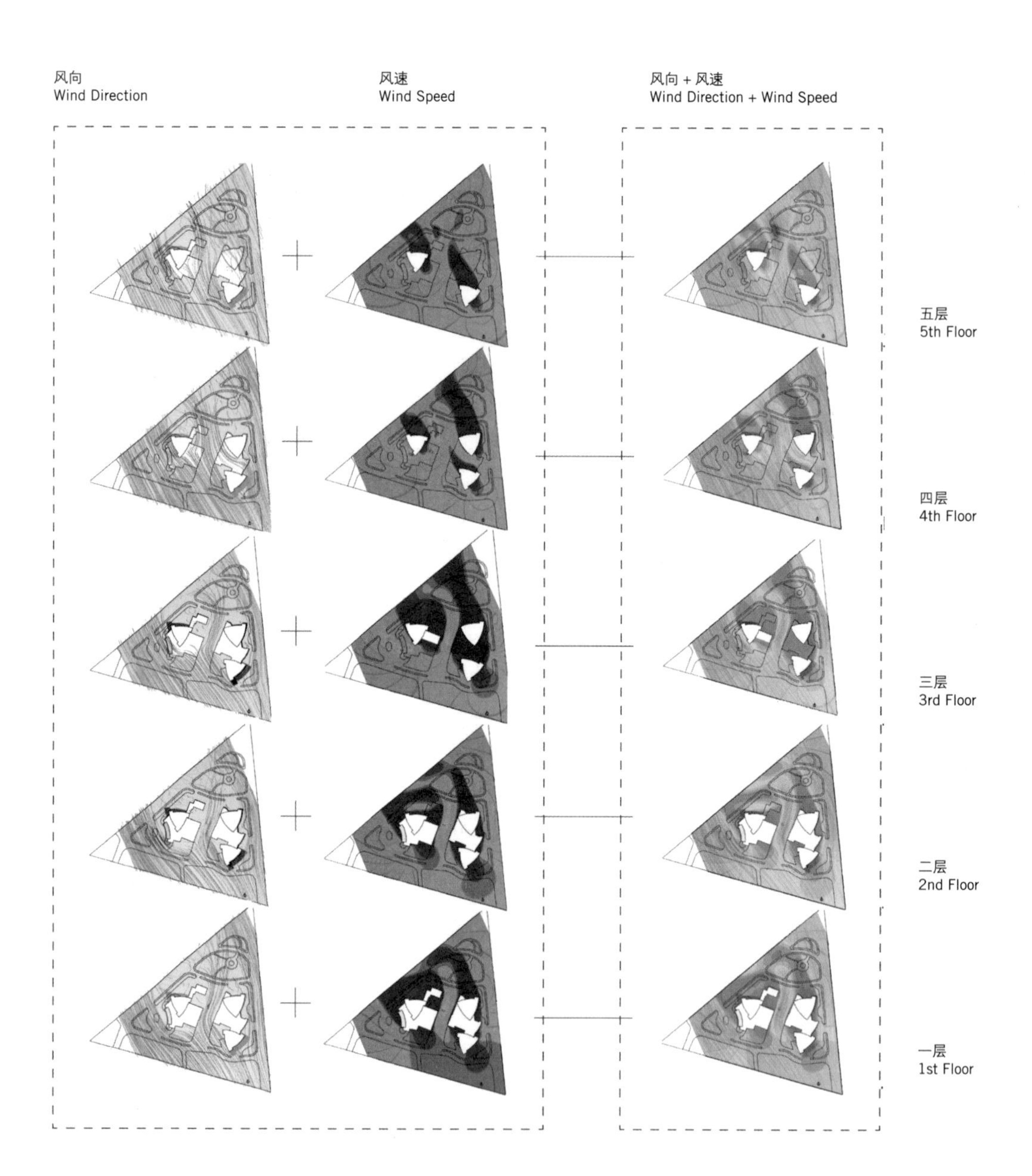


根据风环境模拟的各层孔洞原型生成

Prototype Generated through Wind Environment Simulation

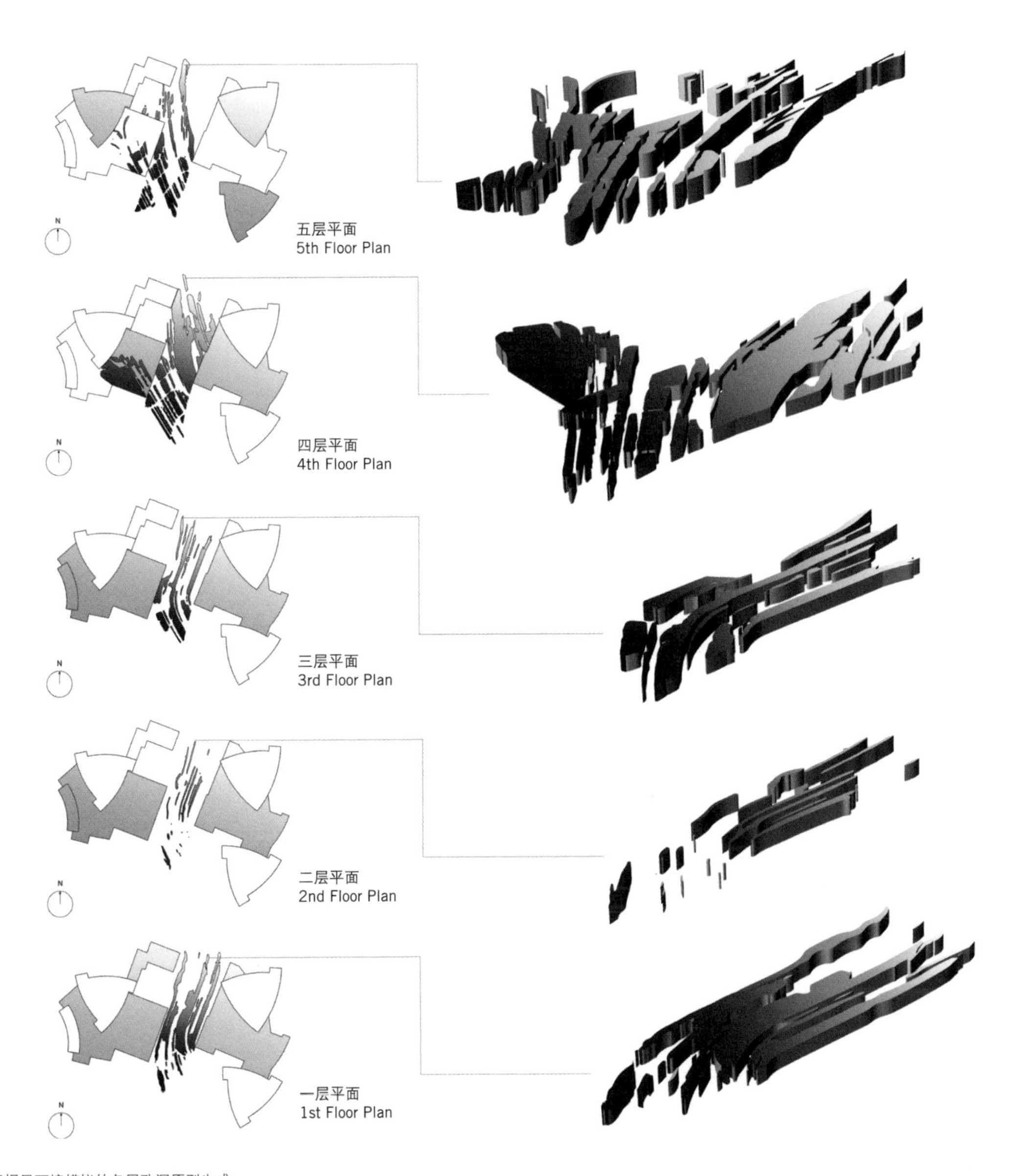


根据风环境模拟的各层孔洞原型生成
Prototype Generated through Wind Environment Simulation

在两幢高层塔楼之间适当拆除和添加，形成一个浮游森林的路径，这既是风的通道，又是激活原有建筑活力的催化空间，建筑拱廊构成的多样化空间是对多西的致敬。

The plan is to properly dismantle and add between the two high-rise towers to form a floating forest path, which is both a passage for the wind and a catalytic space to activate the vitality of the original building. The diverse space of the architectural arcades is a tribute to Doshi.

# 浮游之森
# Gardens on Gardens

学生 Students：
李慧妮 Lee We-Ni
吴剑翎 Kimling Irene Ng Zheng

指导老师 Tutors：
李麟学 LI Linxue
周渐佳 ZHOU Jianjia

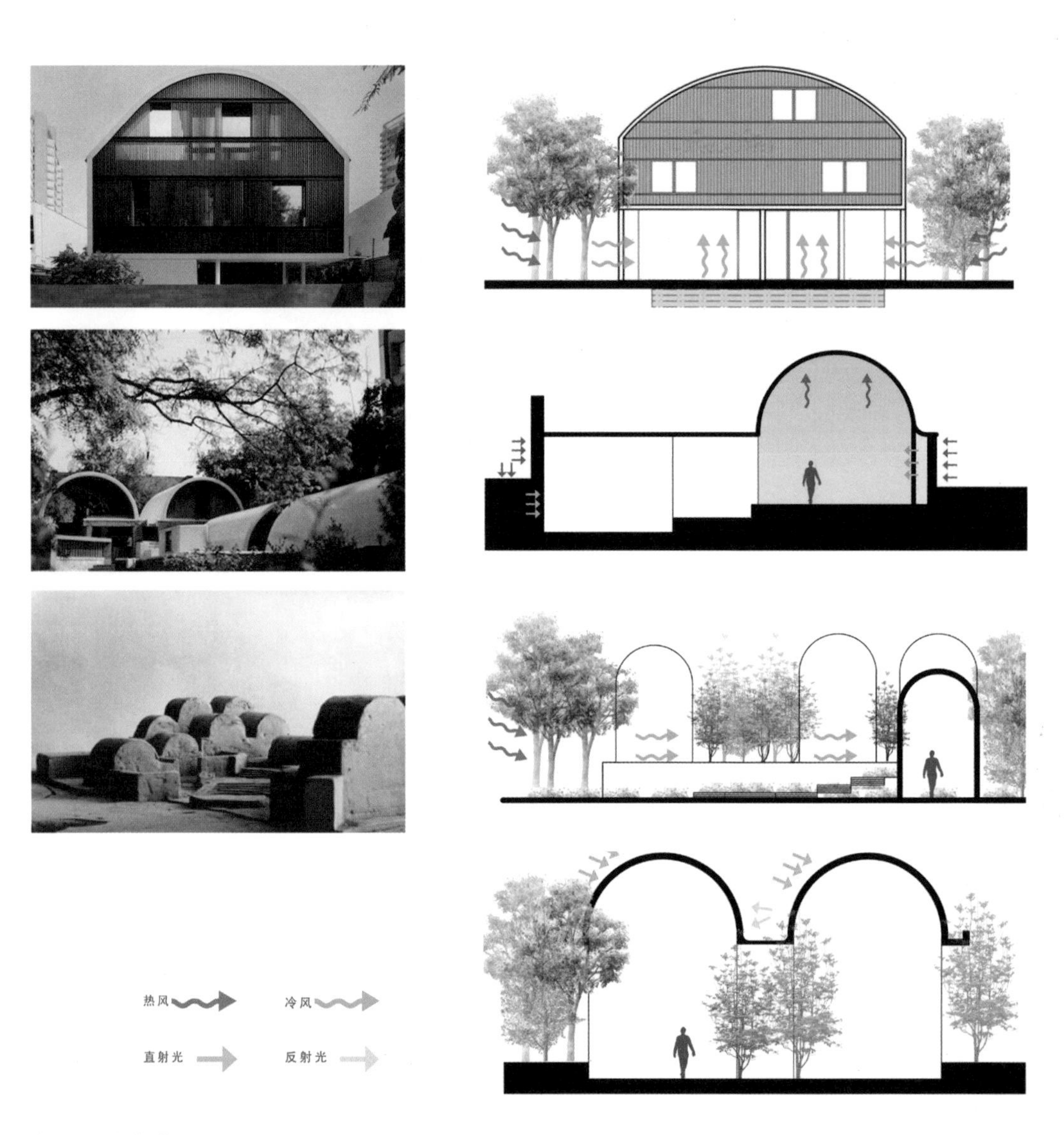


案例研究：新加坡风拱顶住宅
Case Study: Wind Vault House

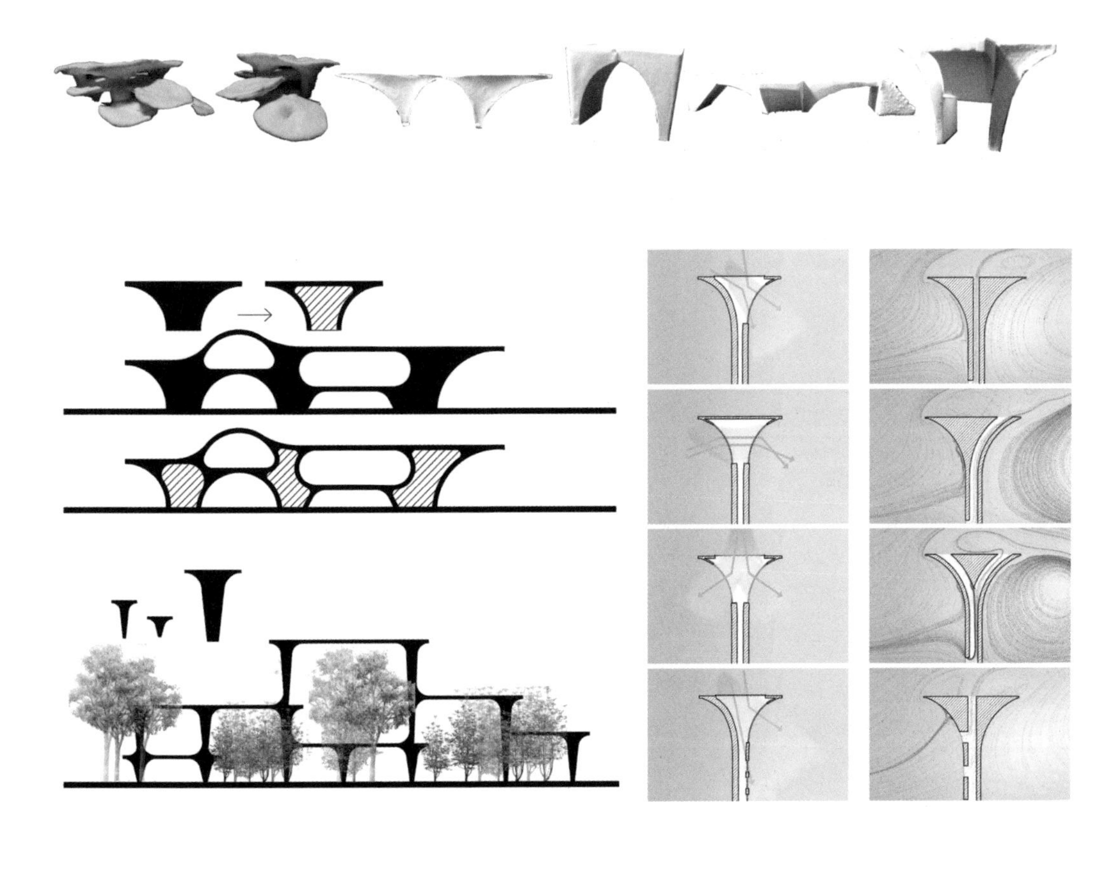

原型研究与原型生成
Prototype Research and Prototype Generation

## 热力学建筑原型 Thermodynamic Architectural Prototype

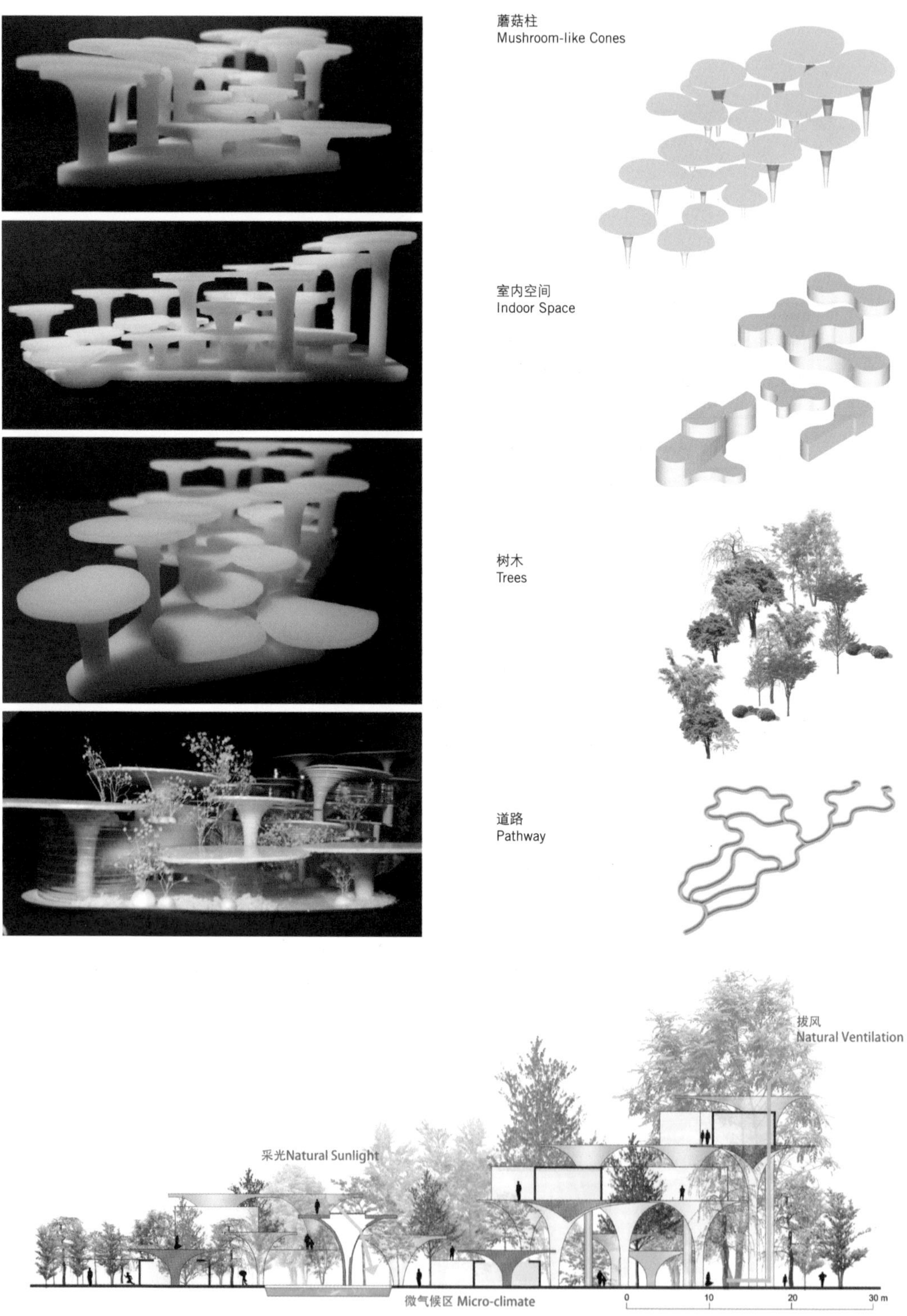


原型组合与植入

Prototype Combination and Implantation

设计以研究广州的“西关大屋”开始，提炼出促进建筑中空气流通的手段，包括：贯通的冷巷加强通风；坡顶、天窗与天井形成有效的导风系统；兼顾热压通风与风压通风。设计以此系统为参照，结合上海的气候与场地环境，以剖面设计和空气流通作为建筑组织的出发点，提供一个根据季节调整开放度的建筑模型。

The design begins with the study of the Xiguan Big House in Guangzhou and extracts the means to promote air circulation in the building, including through the cold alley to enhance ventilation; the top of the slope, the skylight and the patio form an effective air guiding system; taking into account the hot pressure ventilation and wind pressure ventilation. The design is based on this system, combined with Shanghai's climate and site environment, with section design and air circulation as the starting point of the building organization, providing an architectural model that adjusts the openness according to the season.

# 风来风往
# In and Out

学生 Students：
周姝 ZHOU Shu
吴剑翎 Kimling Irene Ng Zheng

指导老师 Tutors：
李麟学 LI Linxue
周渐佳 ZHOU Jianjia

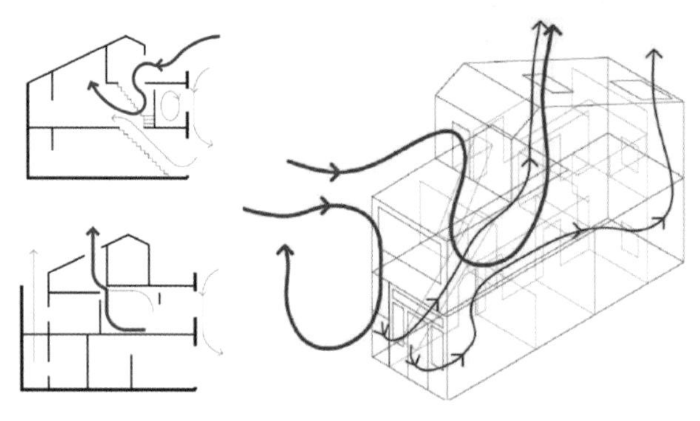

坡顶 + 天窗 + 天井导风系统
Slope Roof + Skylight + Air Guide System

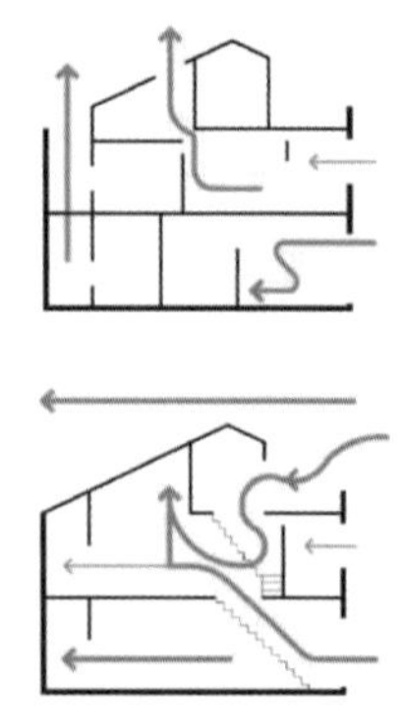
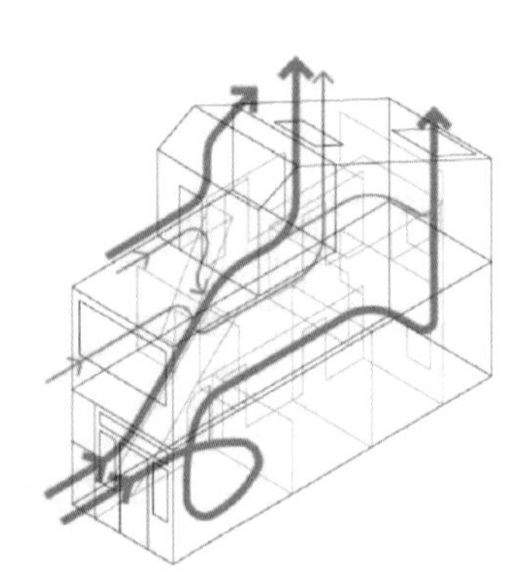

热压 + 风压通风
Thermal Pressure + Wind Pressure Ventilation

案例研究：西关大屋
Case Study: Xiguan Big House

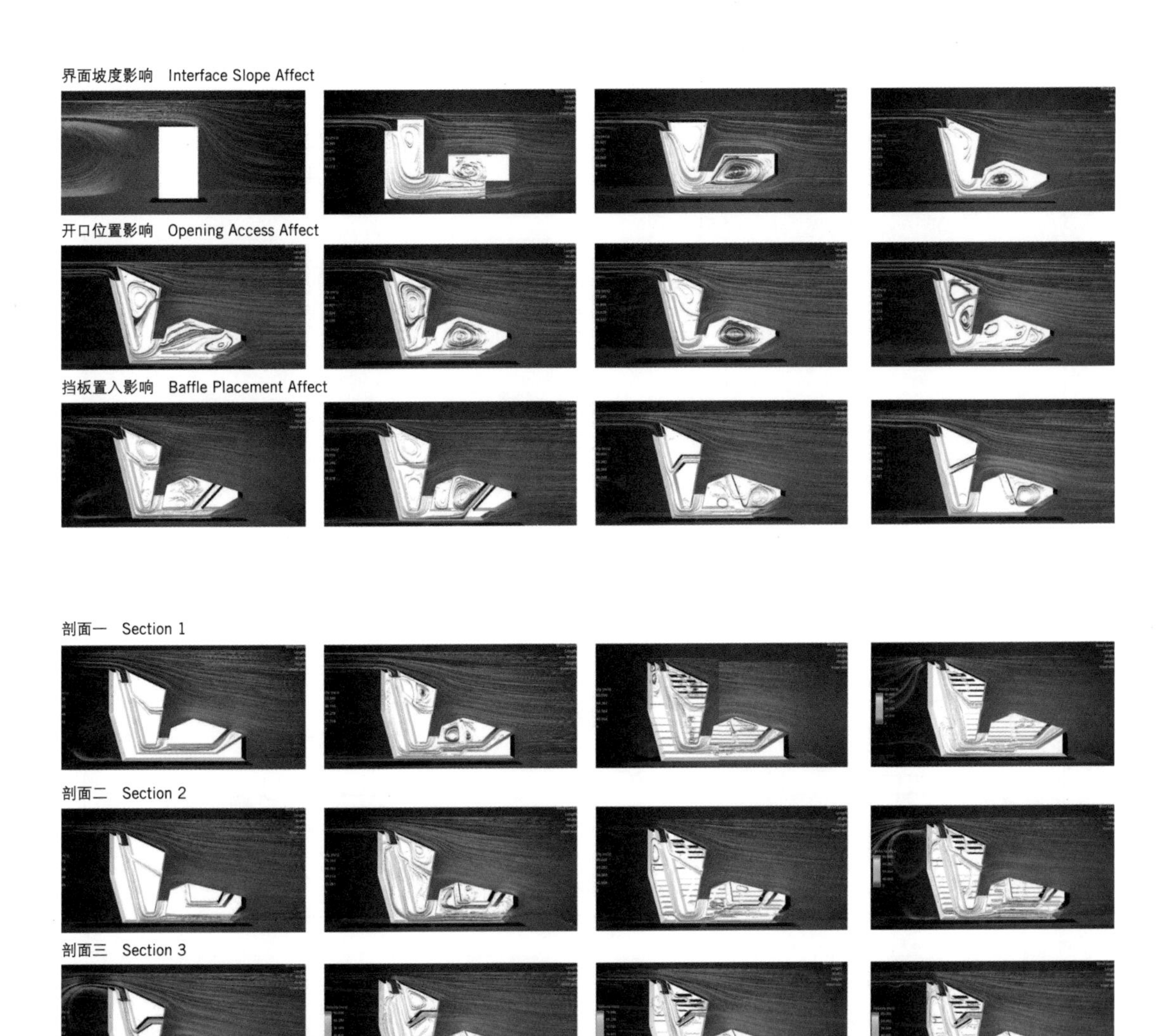


风环境模拟与形态影响要素
Wind Environment Simulation and Factors of Morphological Influence

## 热力学建筑原型 Thermodynamic Architectural Prototype

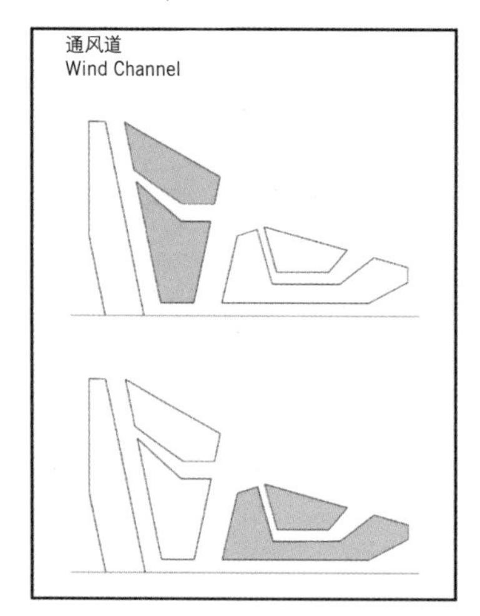


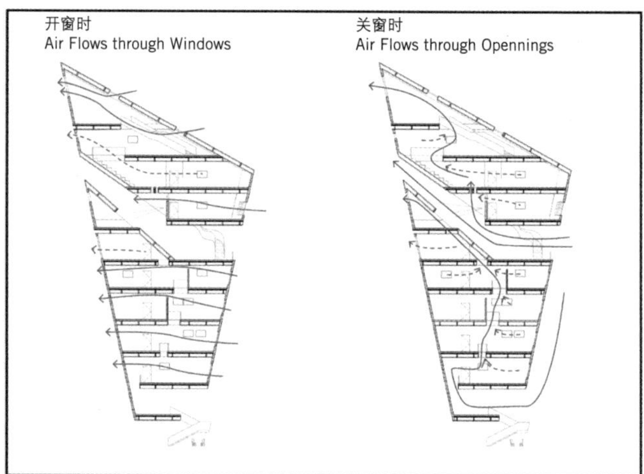


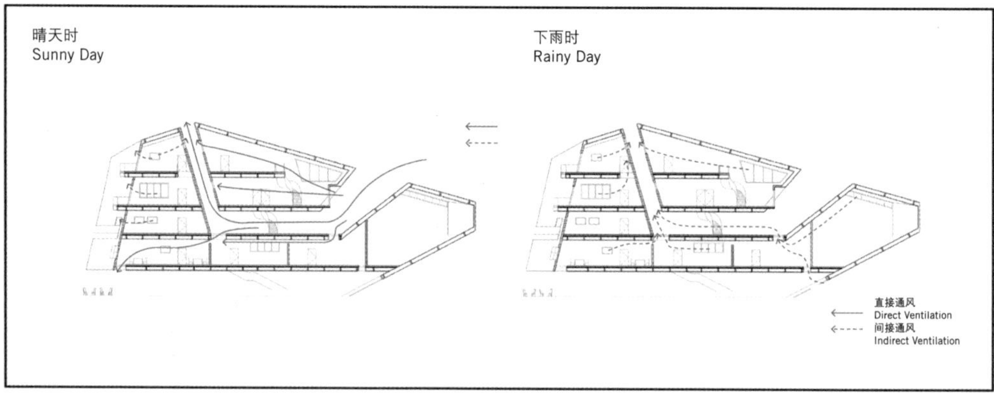


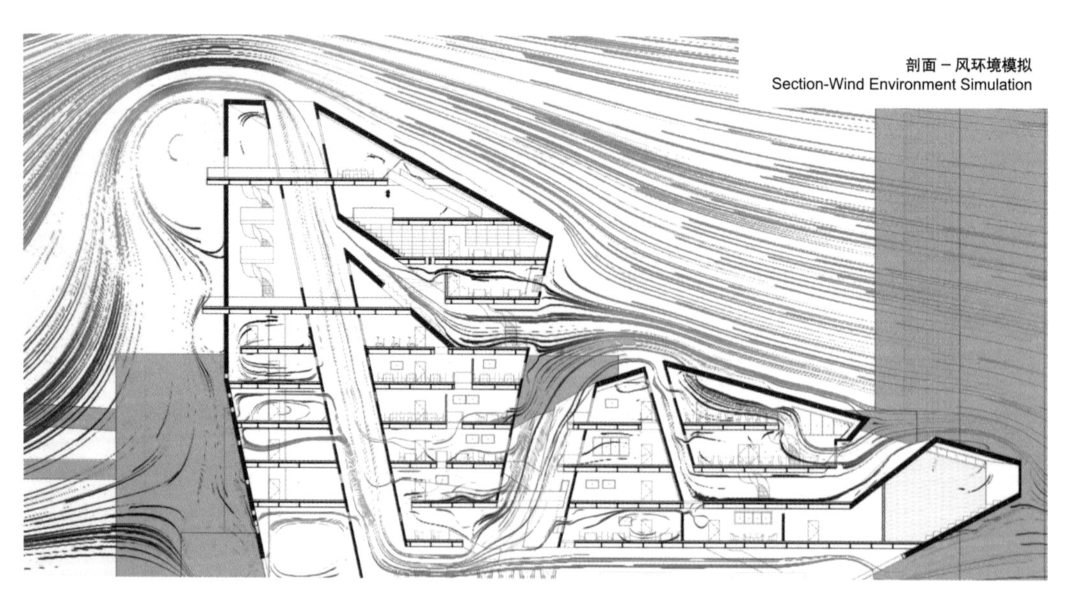


风环境模拟与剖面生成

Wind Environment Simulation and Section Generation

# 光之图书馆
# Library of Light

基地位于上海浦东新区，在世纪大道的尽端，基地西北侧为浦东新区行政办公中心；东侧是浦东新区展览馆和银联大厦；南侧为公共绿地，隔锦绣路是世纪公园。场地城市密度极低，城市尺度巨大，而且绿树成荫。

At the end of Century Avenue, the northwest side of the base is the Administrative Office Center of Pudong New District, the east side is Pudong New District Exhibition Hall and UnionPay Building; the south side is public green space, and Century Park is just across the Jinxiu Road. The density of the area is extremcly low, while the city scale is huge with tree-lined streets.

指导老师：李麟学、周渐佳
评审老师：王骏阳、王方戟、窦平平、李卓、郝洛西、崔哲
助教：黄潇颖、陶思旻
学生：郑思尧、郑馨、吕欣欣、林静之、杨千惠、王劲凯、许双盈、张若松、王宇昊
Tutors: LI Linxue, ZHOU Jianjia
Review Teachers: WANG Junyang, WANG Fangji, Dou Pingping, LI Zhuo, HAO Luoxi, CUI Zhe
Teaching Assistants: HUANG Xiaoying, TAO Simin
Students: ZHENG Siyao, ZHENG Xin, LYU Xinxin, LIN Jingzhi, YANG Qianhui, WANG Jinkai, XU Shuangying, ZHANG Ruosong, WANG Yuhao

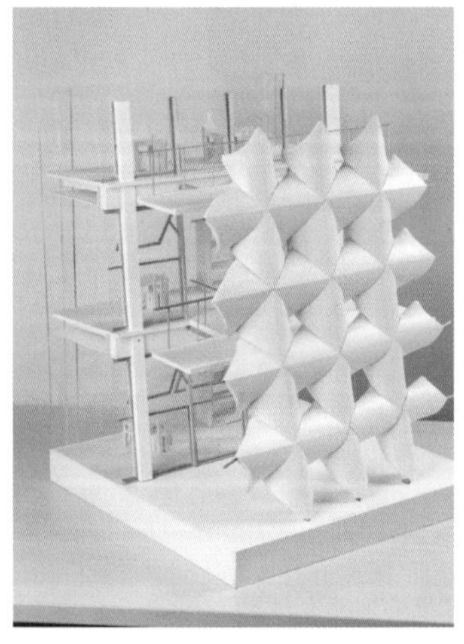

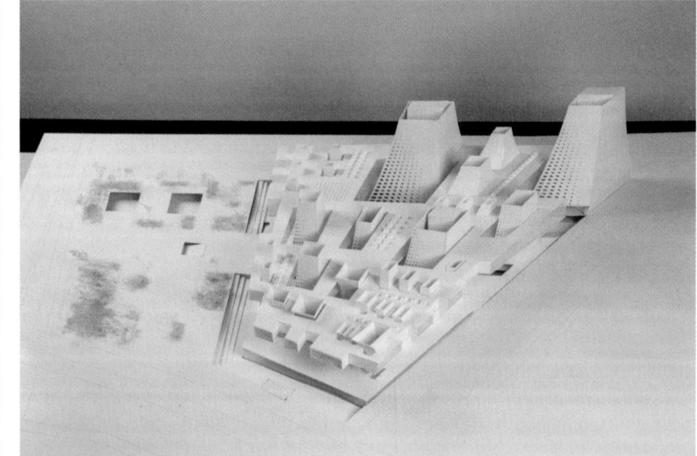

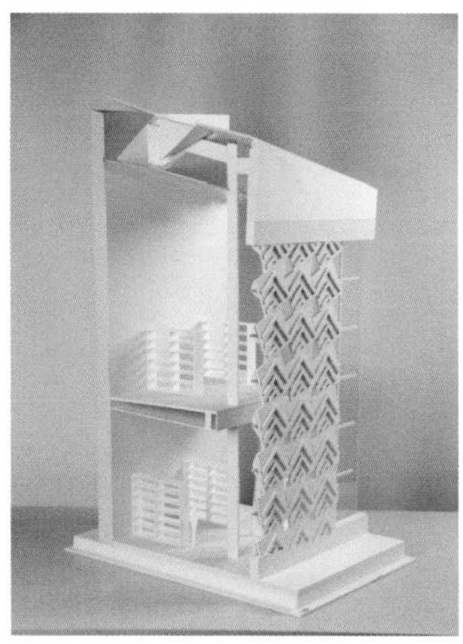

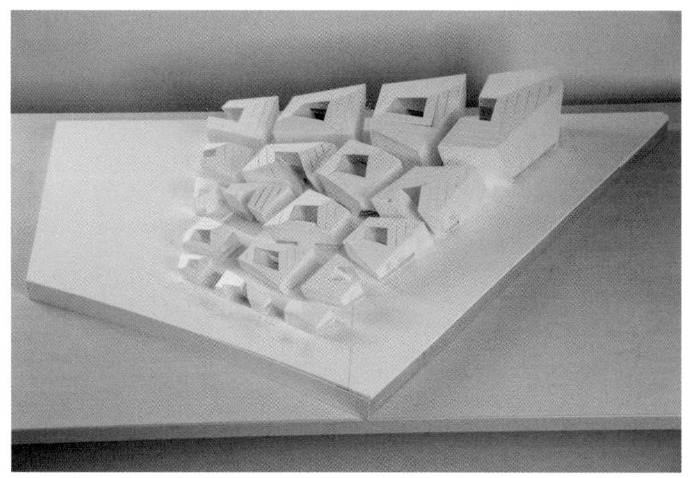

左页：光合作用模型照片
右页：光的聚落模型照片
Left Page: Model photo of Photosynthesis
Right Page: Model photo of Interrelated Courtyard

我们将树木的能量机制作为研究光和能量的切入点，以树木的能量机制中孔隙和层次的因素为重点进行研究，“光合作用”这个题目也就此诞生。经过研究后，确定以孔隙、通道和分层作为关键词，将树对能量的利用划分为六个机制，分别是捕捉、转化、传导、传输、储存和输出，将此能量机制运用到建筑上生成一个上层有开孔、下层风塔状，且上下层交叠的通道。我们尝试了一系列原型模型来研究三个关键词对光的影响，并基于原型研究中所确立的基本准则对不同尺度的原型进行组合，从而生成建筑的基本形态。再通过光环境和风环境的模拟进行优化，得到最终的建筑形态，实现了六个机制在建筑中的运行。

We took the energy mechanism of trees as the starting point to study light and energy. Focus on the factors of porosity and layer in the energy mechanism trees gave us the title of this proposal: Photosynthesis. After study, we divided the energy mechanism of trees into six sections i.e. capture, conversion, transfer, transmission, storage and output. We applied it to build a prototype with pores on the top layer and a wind-tower type channel on the bottom layer. We fabricated a number of models with varied pore size, channel form and layer thickness to study the influence of the three keywords on lighting. We further defined the prototype in architecture by gradual deduction of light, wind and public space for different layers. Through simulation of wind and light environment, the final form of architecture was generated, realizing the natural mechanism into architecture.

# 光合作用
# Photosynthesis

学生 Students：
郑馨 ZHENG Xin
郑思尧 ZHENG Siyao
吕欣欣 LYU Xinxin

指导老师 Tutors：
李麟学 LI Linxue
周渐佳 ZHOU Jianjia

## 热力学建筑原型 Thermodynamic Architectural Prototype

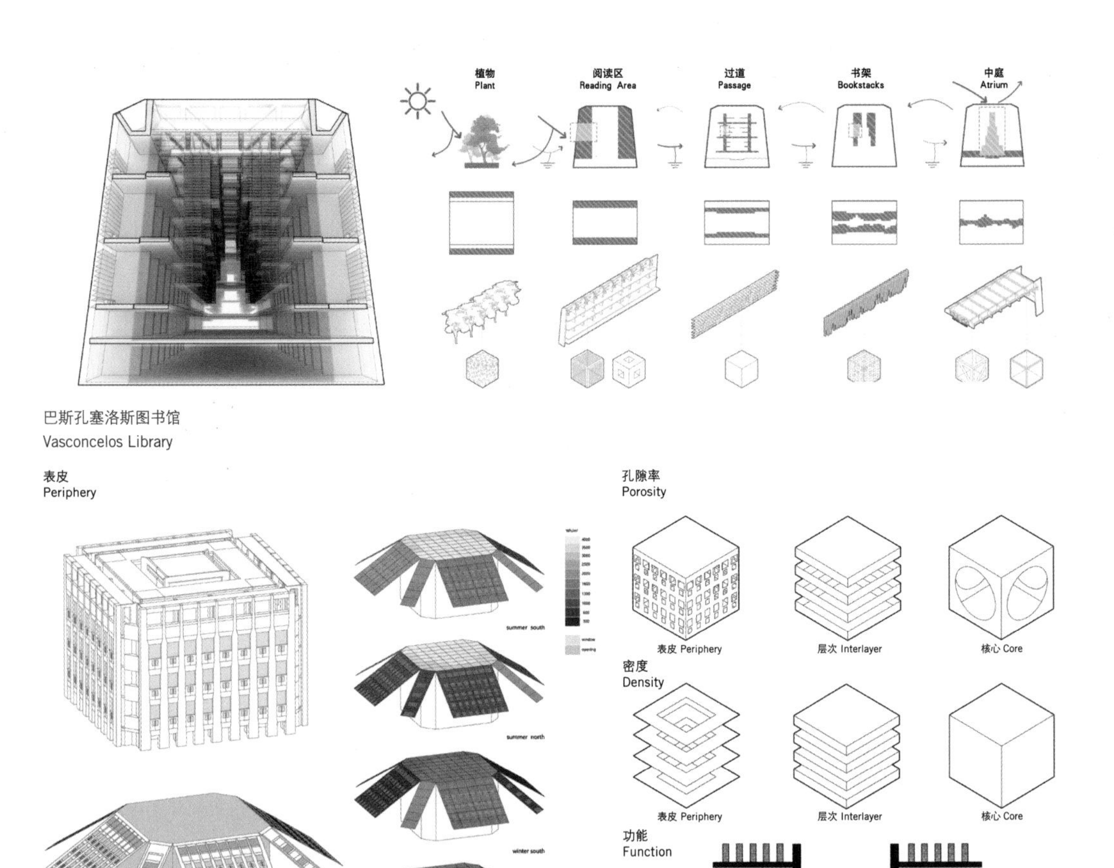


巴斯孔塞洛斯图书馆
Vasconcelos Library

菲利普斯埃克塞特学院图书馆
Phillips Exeter Academy Library

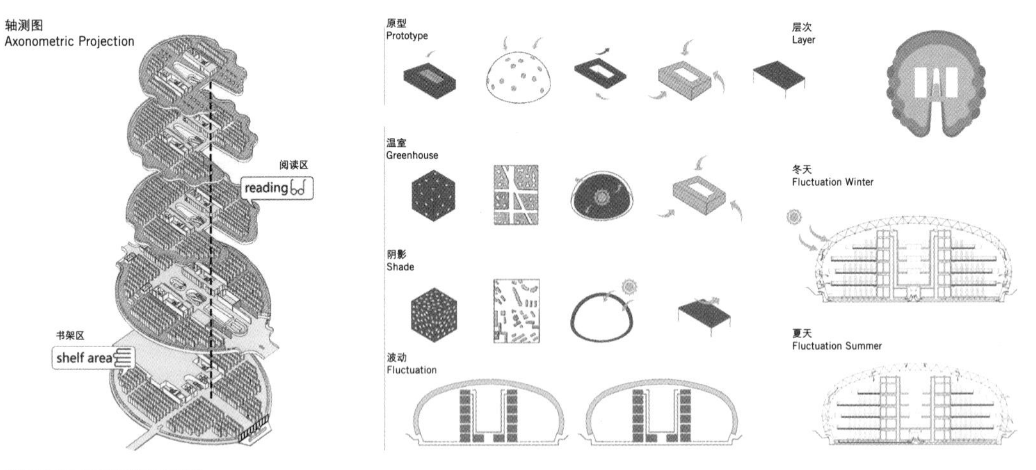


柏林自由大学语言学院图书馆
Philological Library, Free University of Berlin

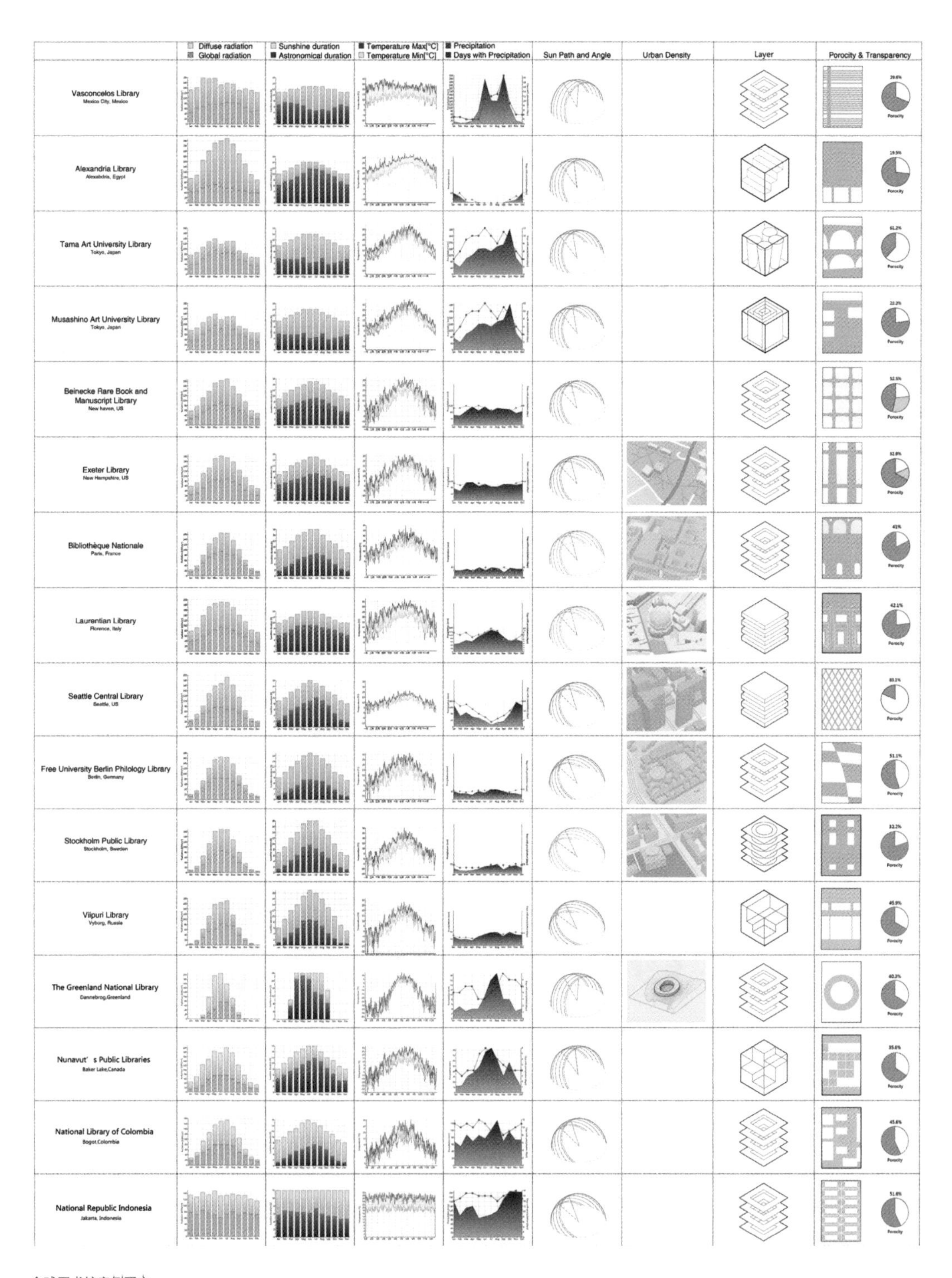
Diffuse radiation
Global radiation
Sunshine duration
Astronomical duration
Temperature Max[°C]
Temperature Min[°C]
Precipitation
Days with Precipitation
Sun Path and Angle
Urban Density
Layer
Porocity & Transparency
Vasconcelos Library
Mexico City, Mexico
Alexandria Library
Alexabdria, Egypt
Tama Art University Library
Tokyo, Japan
Musashino Art University Library
Tokyo, Japan
Beinecke Rare Book and
Manuscript Library
New haven, US
Exeter Library
New Hampshire, US
Bibliothèque Nationale
Paris, France
Laurentian Library
Florence, Italy
Seattle Central Library
Seattle, US
Free University Berlin Philology Library
Berlin, Germany
Stockholm Public Library
Stockholm, Sweden
Viipuri Library
Vyborg, Russia
The Greenland National Library
Dannebrog,Greenland
Nunavut' s Public Libraries
Baker Lake,Canada
National Library of Colombia
Bogot,Colombia
National Republic Indonesia
Jakarta, Indonesia

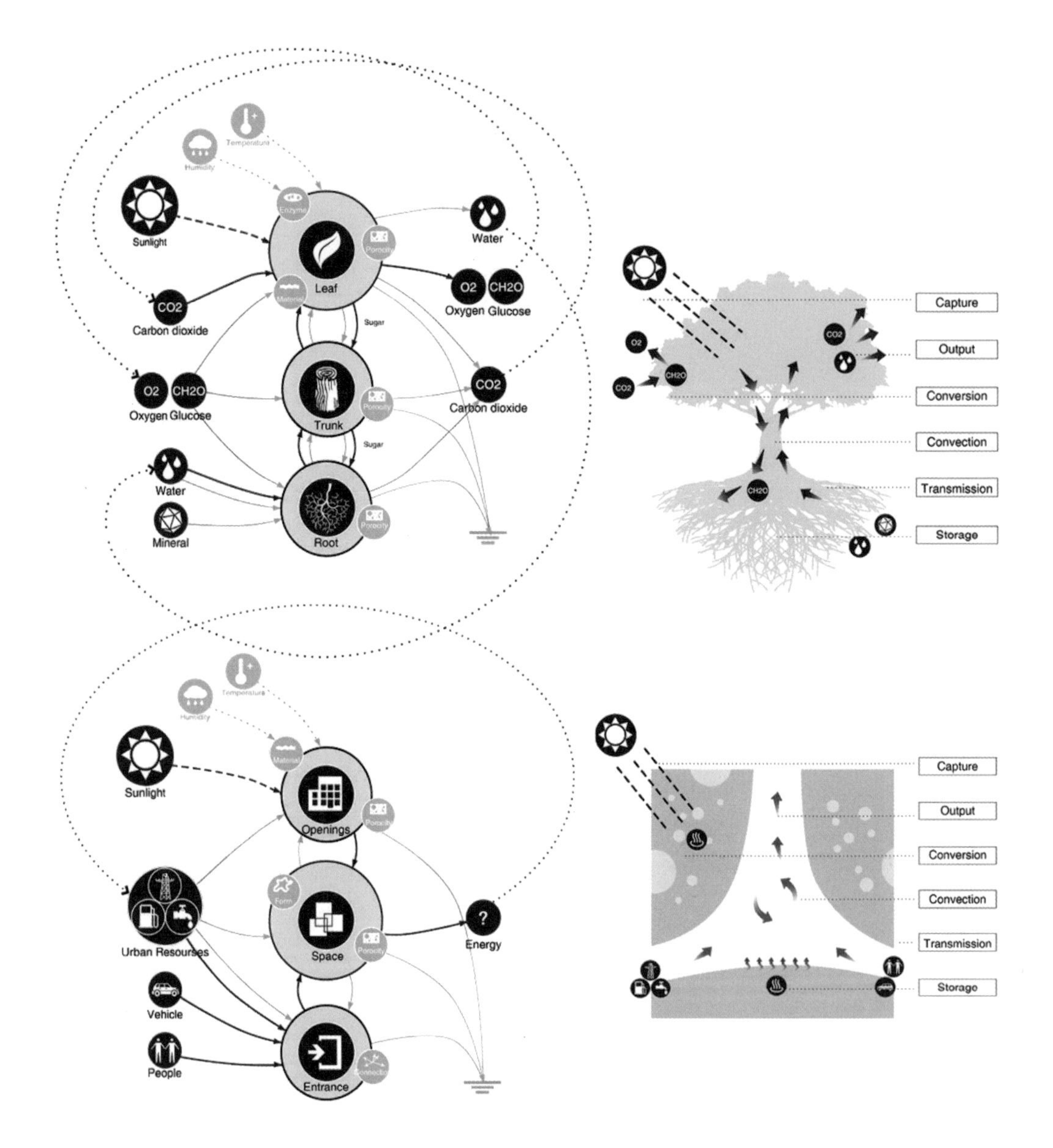


能量流动与光合作用

Energy Flow and Photosynthesis Principle

植物光合作用的原理可分为六个过程。通过这些过程，植物可以将大量的太阳能转移到其生长中，并在其呼吸和蒸腾过程中重复利用物质和能量，使能量尽可能地流动和循环。

The principle of photosynthesis in plants include six steps through which plants could transfer a large amount of solar energy to its growth and reuse the matter and energy in its respiration and transpiration, making the energy flowing and circulating as far as possible.

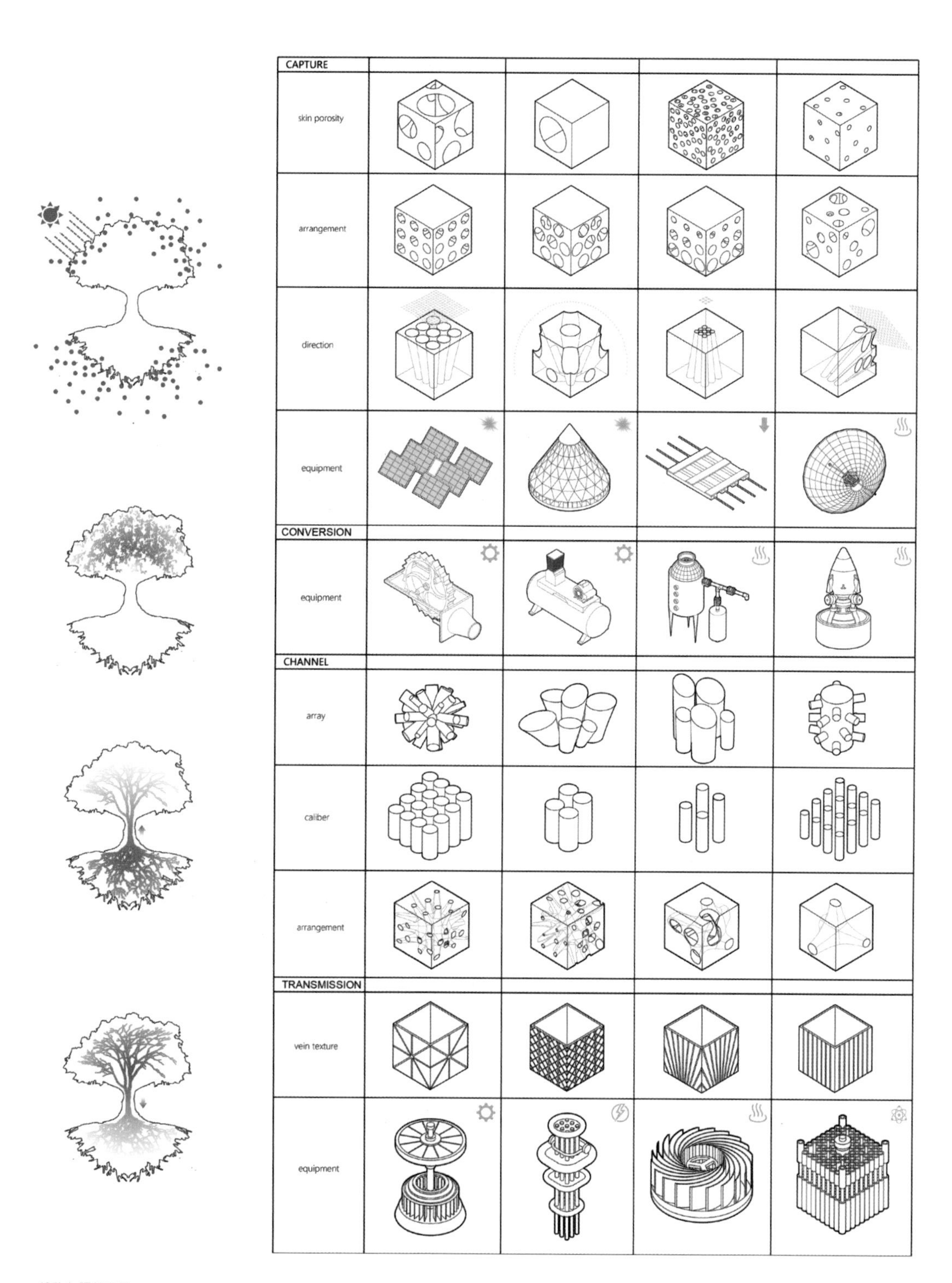


植物与建筑要素
Analogy of Plants and Building Elements

## 热力学建筑原型 Thermodynamic Architectural Prototype

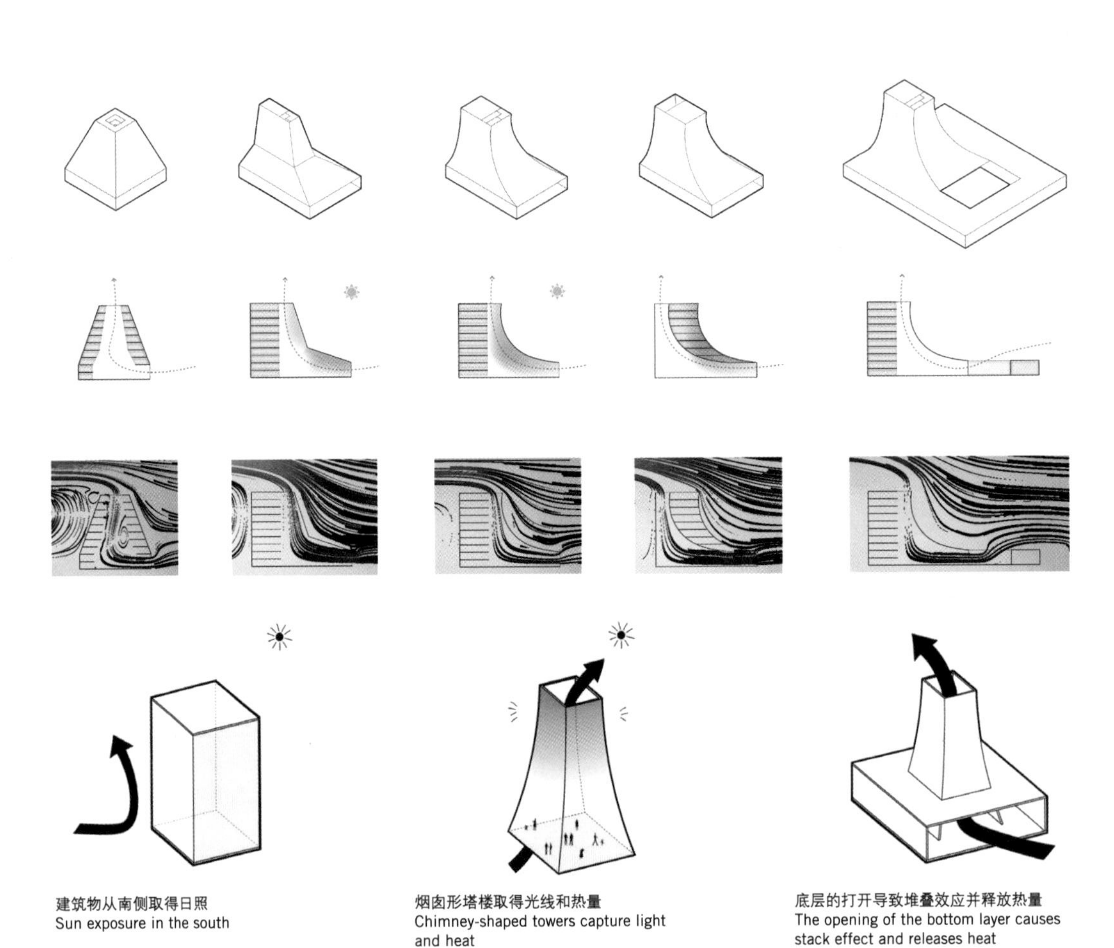

建筑物从南侧取得日照
Sun exposure in the south

烟囱形塔楼取得光线和热量
Chimney-shaped towers capture light and heat

底层的打开导致堆叠效应并释放热量
The opening of the bottom layer causes stack effect and releases heat

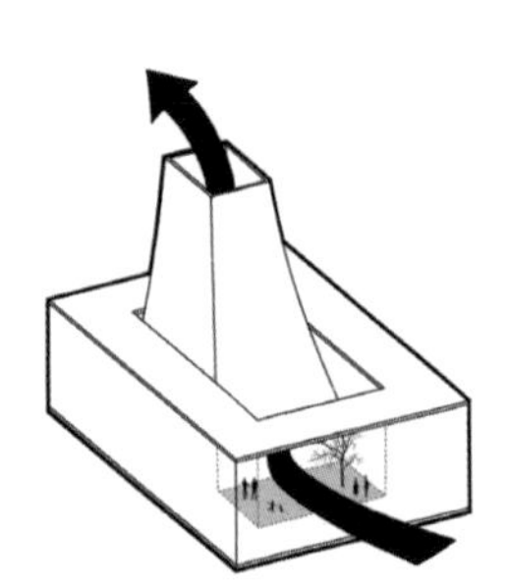

releases heat.

将塔向北倾斜，从而形成花园和更好的通风
Lean the towers to the north side in order to form gardens and better ventilation

抬升一层，形成一个中间层，作为城市公园并得到不同程度的照明
Uplift a layer to create an interlayer which acts as city park and cause different lighting

将穿孔材料应用于正面，从而创造更好的采光和通风
Apply the perforated material to the facade in order to create better daylighting and ventilation

原型生成
Prototype Generation

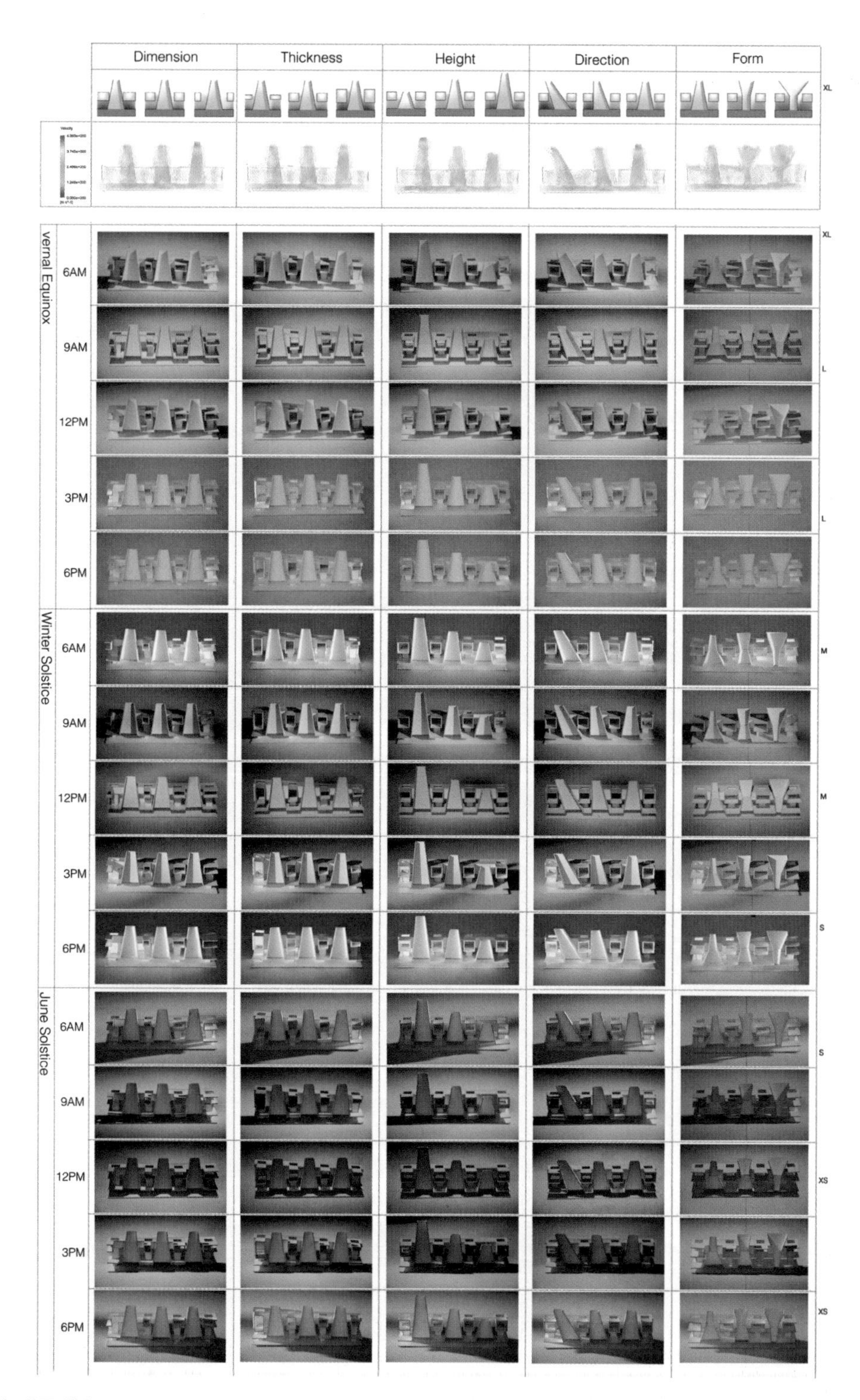


光环境实验下的原型优化
Prototype Optimization in Sunlight Simulation Laboratory

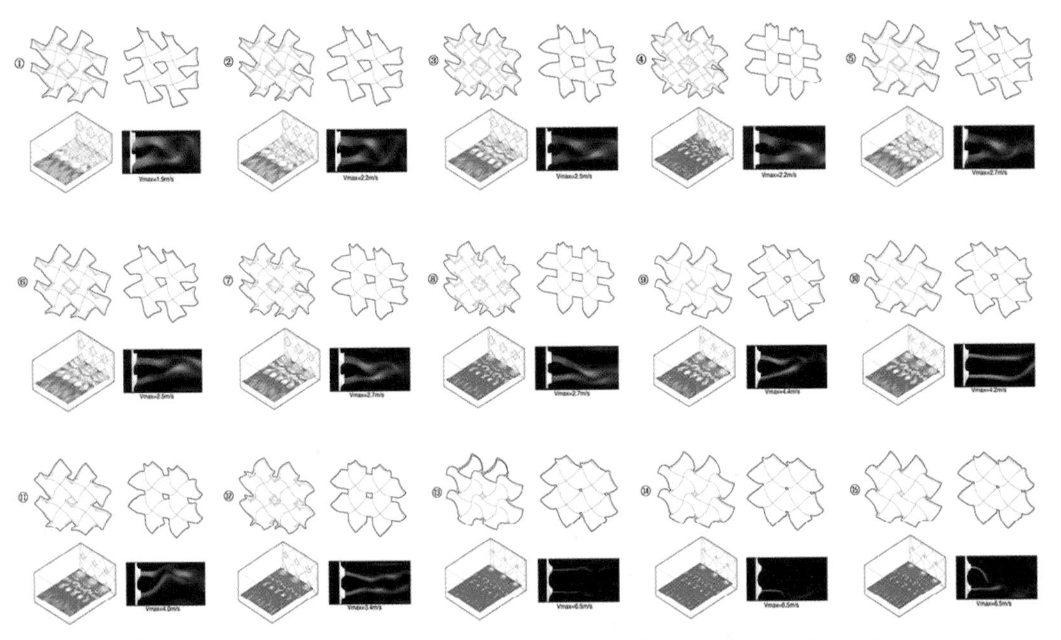

### 幕墙构件的计算机模拟

幕墙构建的设计目的是在一个从下至上的微环境中控制光和风。扭转的程度可以指导下一层的风流并阻止上部风的产生。开口的大小可以稍微控制日光

### Computer Simulation for Facade Components

The component of facade was designed to control the light and wind in a micro-environment from bottom to top. The degree of twisting could guide the wind-flow in the lower floor and block the wind out on upper side. And the scale of openings could slightly control the daylight

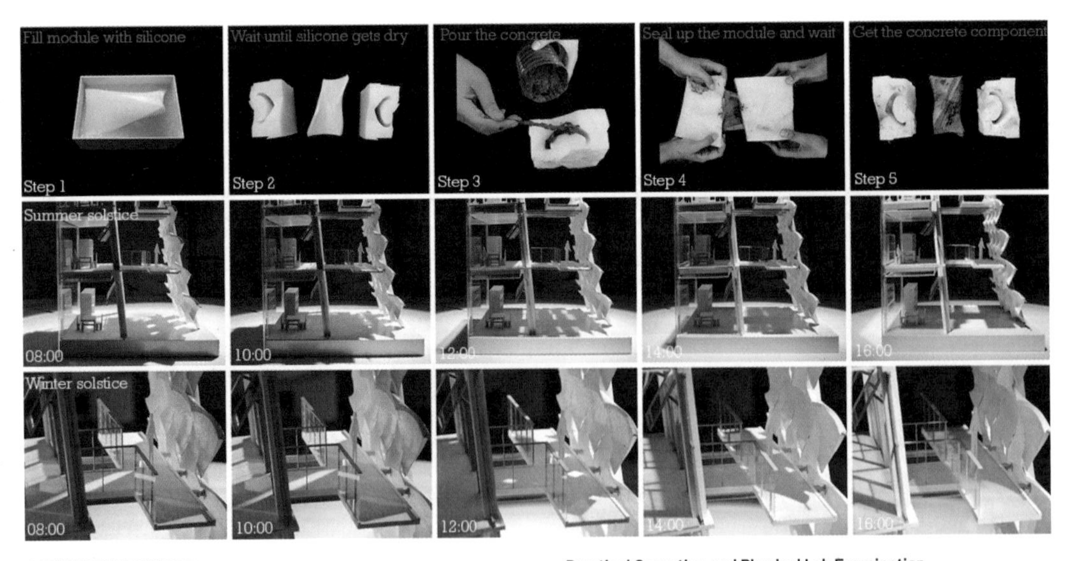


### 实际操作和物理实验检查

所有这些部件都可以用 3D 打印原型和快速硅模制作，证明了其在施工中的可行性。物理氦氖模拟证明了它的热力学性能

### Practical Operation and Physical Lab Examination

All of these components could be fabricated by 3d printed prototype and rapid silica mould, which prove its feasibility in construction. The physical heliodon simulation proves its thermodynamic performance

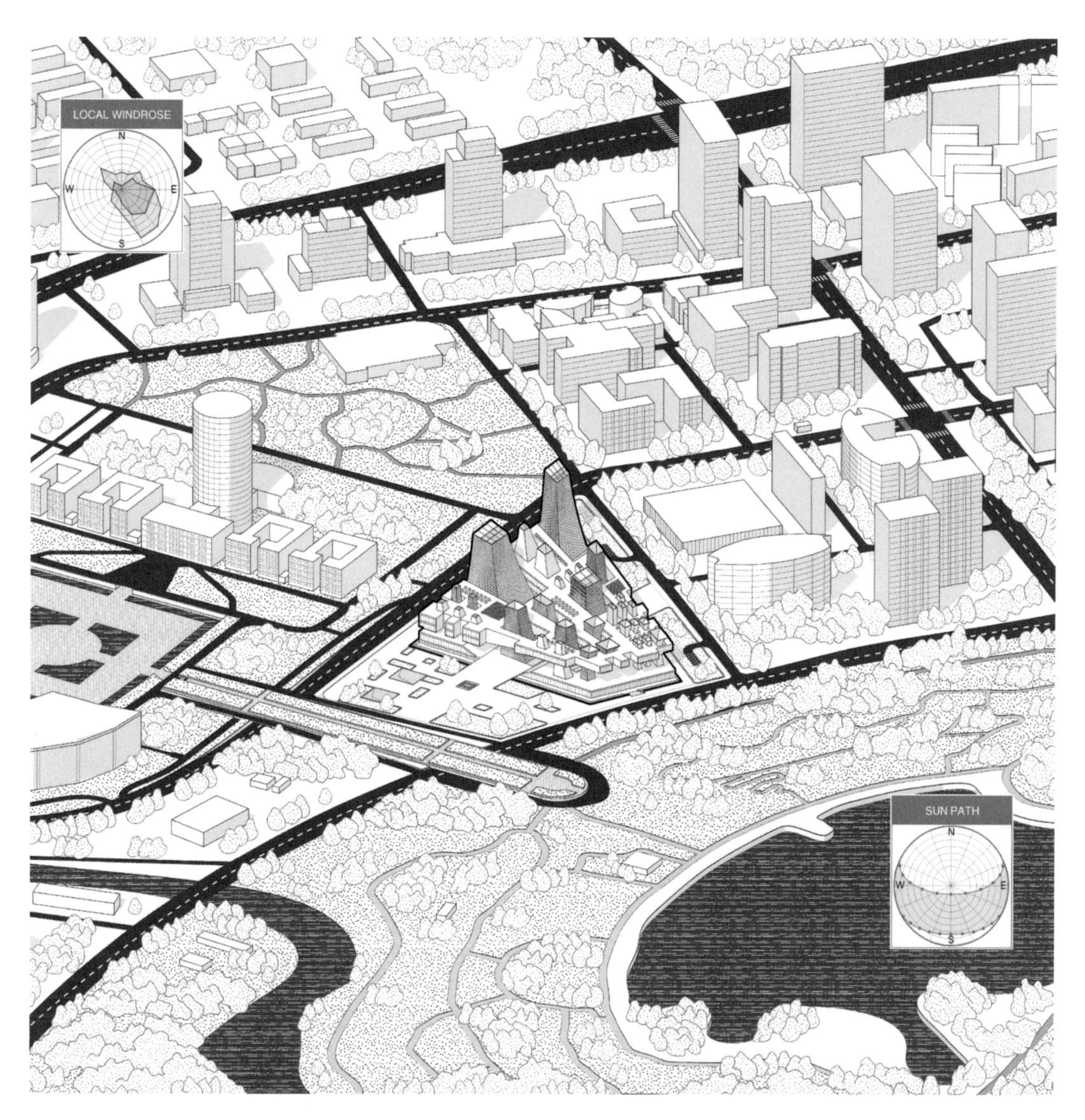

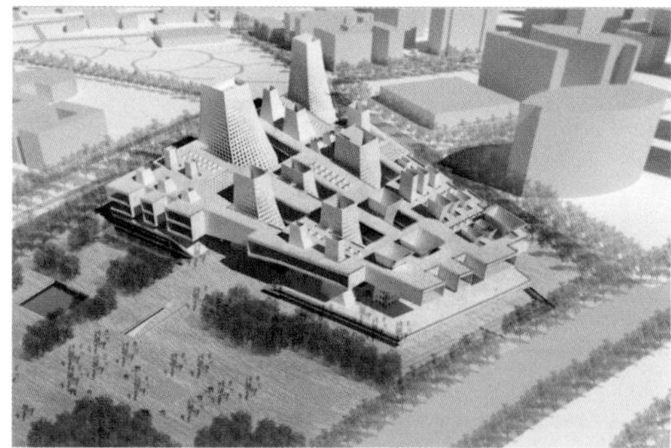

原型植入城市
Prototype Implantation

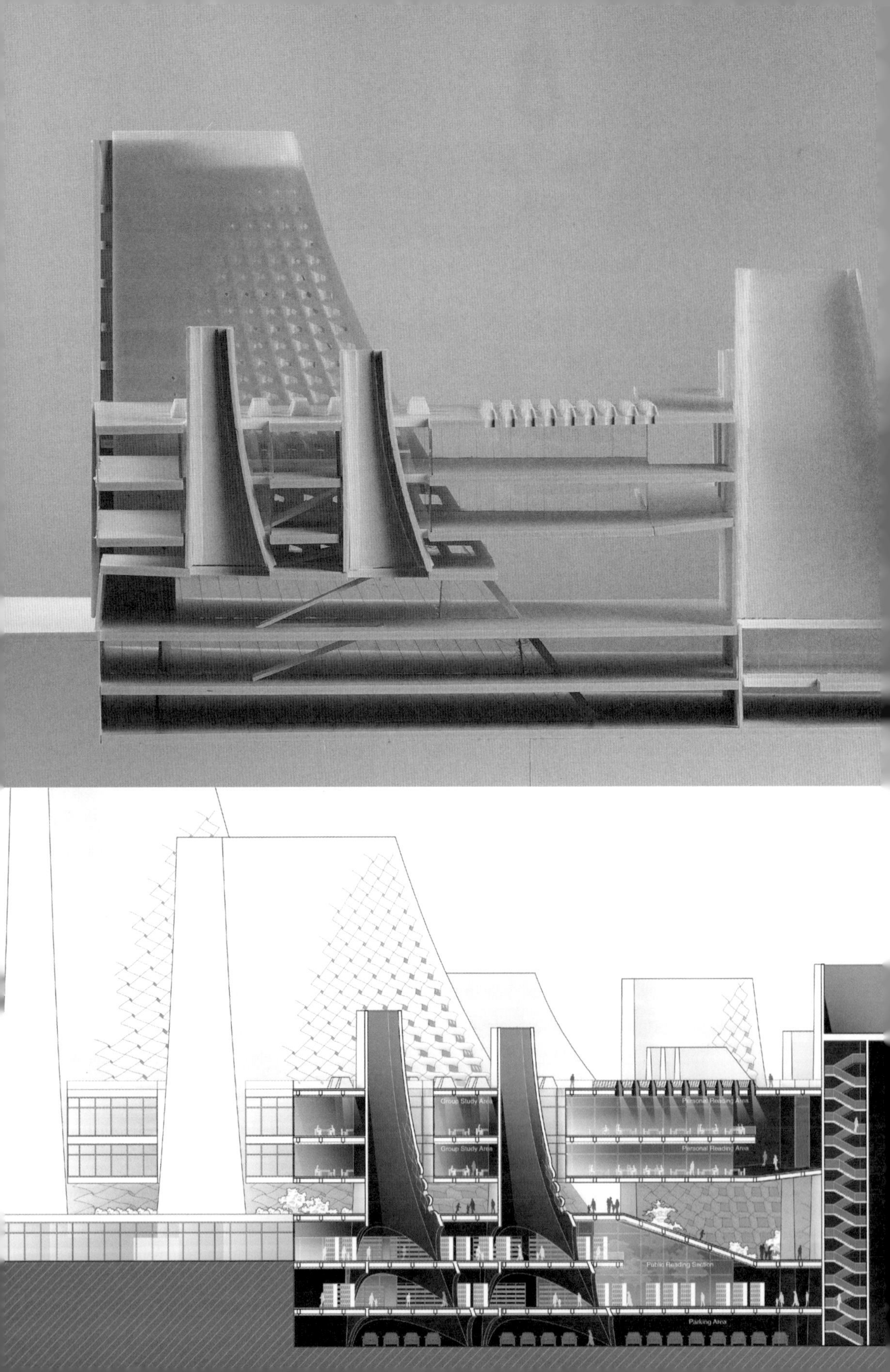
Group Study Area
Personal Reading Area
Group Study Area
Personal Reading Area
Public Reading Section
Parking Area

Roaming Reading Area
Roaming Reading Area
Exhibition Area
Exhibition Area
Equipment Room
Storage Area
Storage Area

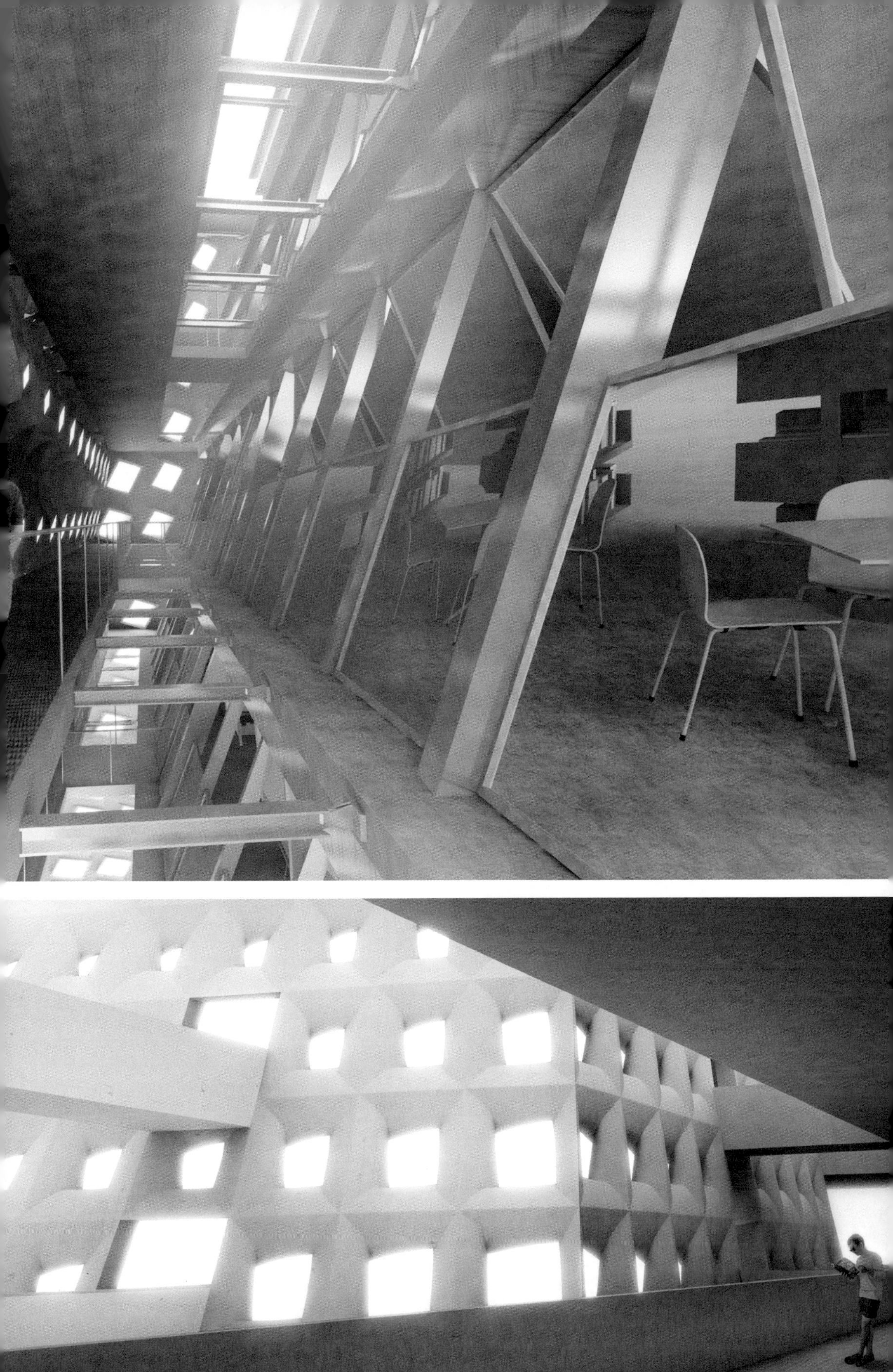

快速发展的上海陆家嘴地区代表了一种夸张的规模，其部分成因受到了该地区建筑物采光条例的影响。然而，建筑物之间的巨大距离导致过度暴露的公共空间和立面，忽略了热舒适性。本方案批评了这种对建筑物之间相互作用的忽视。我们研究了传统民居里蕴藏的简单但有效的气候形态适应性。受到传统民居的启发，这个集群的图书馆旨在探索一种新的图书馆类型，具有舒适的热体验。

The enormous scale of Lujiazui area in Shanghai may be partly due to the sunlight regulations, yet the huge distances between buildings result in overexposure of public spaces and facades, is a complete sacrifice of thermal comfort.
This project critizes the ignorance of interactions between buildings. Inspired by the natural settlements which developed simple but useful morphology adaptations towards climate, the conglomerated library aims to explore new library typology with thermal comfort for snug experiences.

# 光的聚落
# Interrelated Courtyard

学生 Students：
梁芊荟 LIANG Qianhui
林静之 LIN Jingzhi
王劲凯 WANG Jinkai

指导老师 Tutors：
李麟学 LI Linxue
周渐佳 ZHOU Jianjia

**热力学建筑原型 Thermodynamic Architectural Prototype**

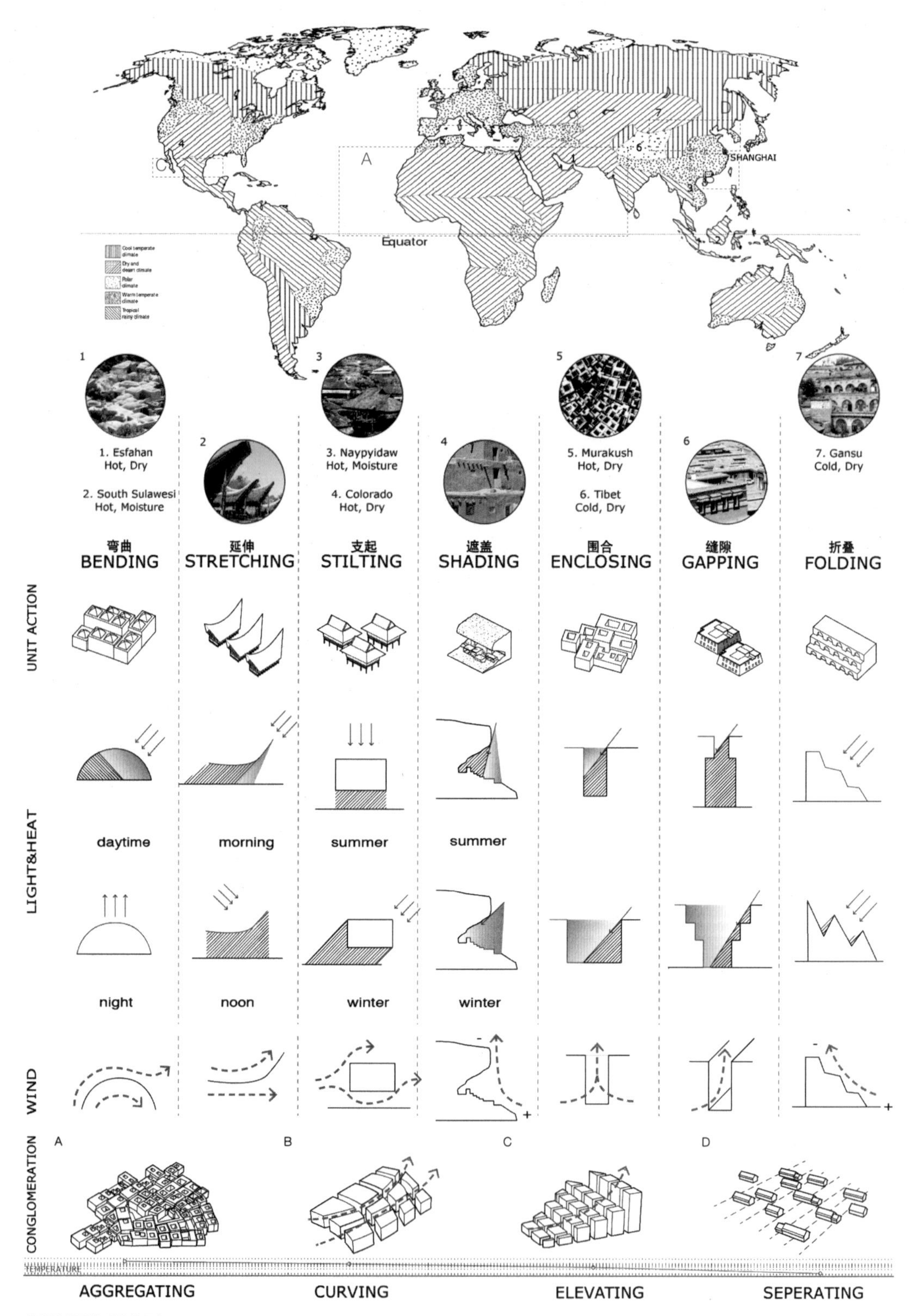


传统民居形态适应性研究

Natural Settlements' Morphology Adaptations

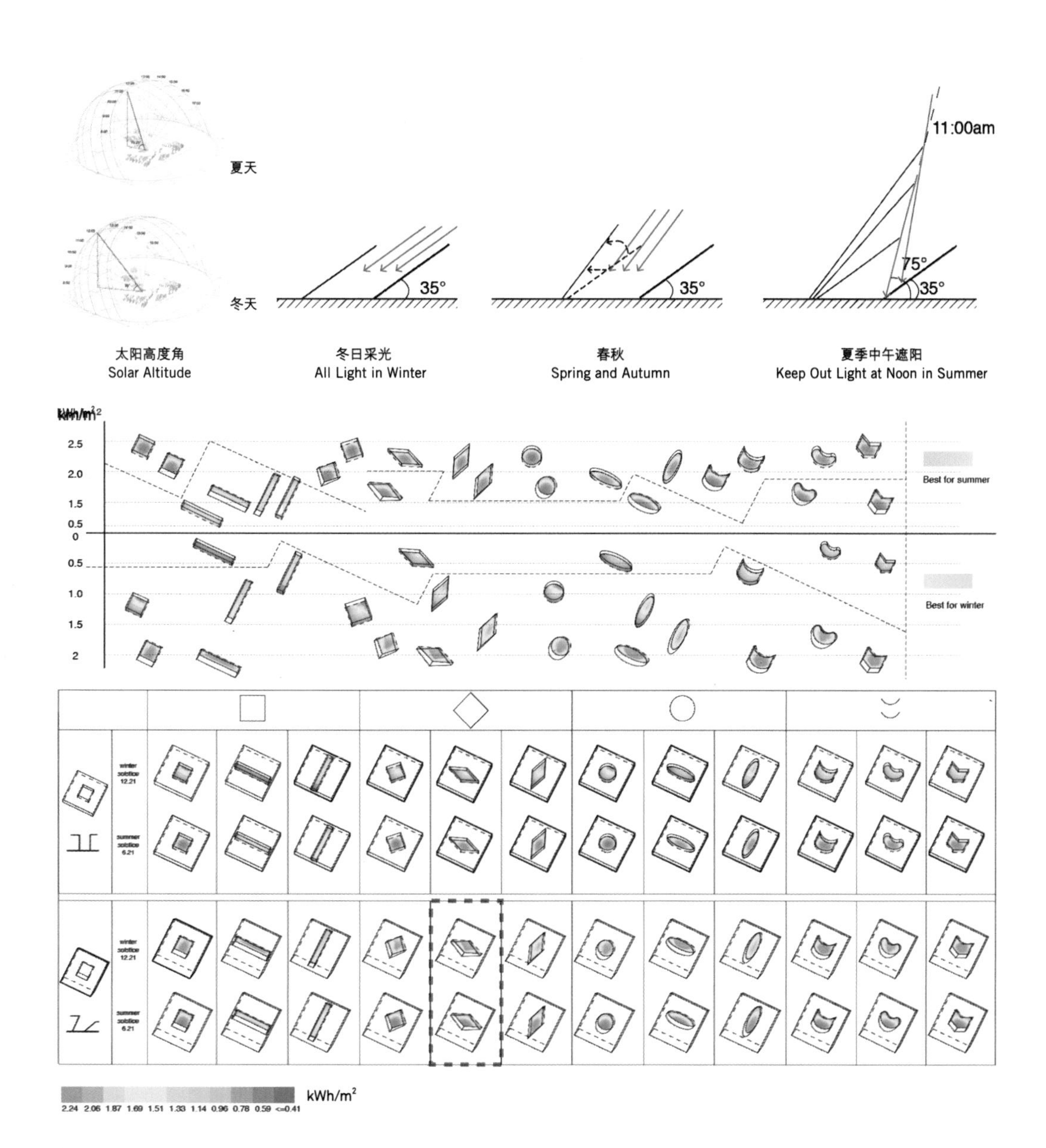


院落变形研究
Courtyard Transformation

在夏季和冬季，人们利用不同的太阳高度角的光策略：在冬季捕捉更多的光线，在夏季阻挡强烈的阳光。传统的长方形庭院被用来优化辐射差异以获得舒适的感觉。

The light strategy is proposed to make use of different solar elevation angles in summer and winter: to capture more light during the winter and to keep out the strong sunlight during the summer. Traditional rectangular courtyards are chanlleged to optimize radiation differences for more comfort.

## 热力学建筑原型 Thermodynamic Architectural Prototype

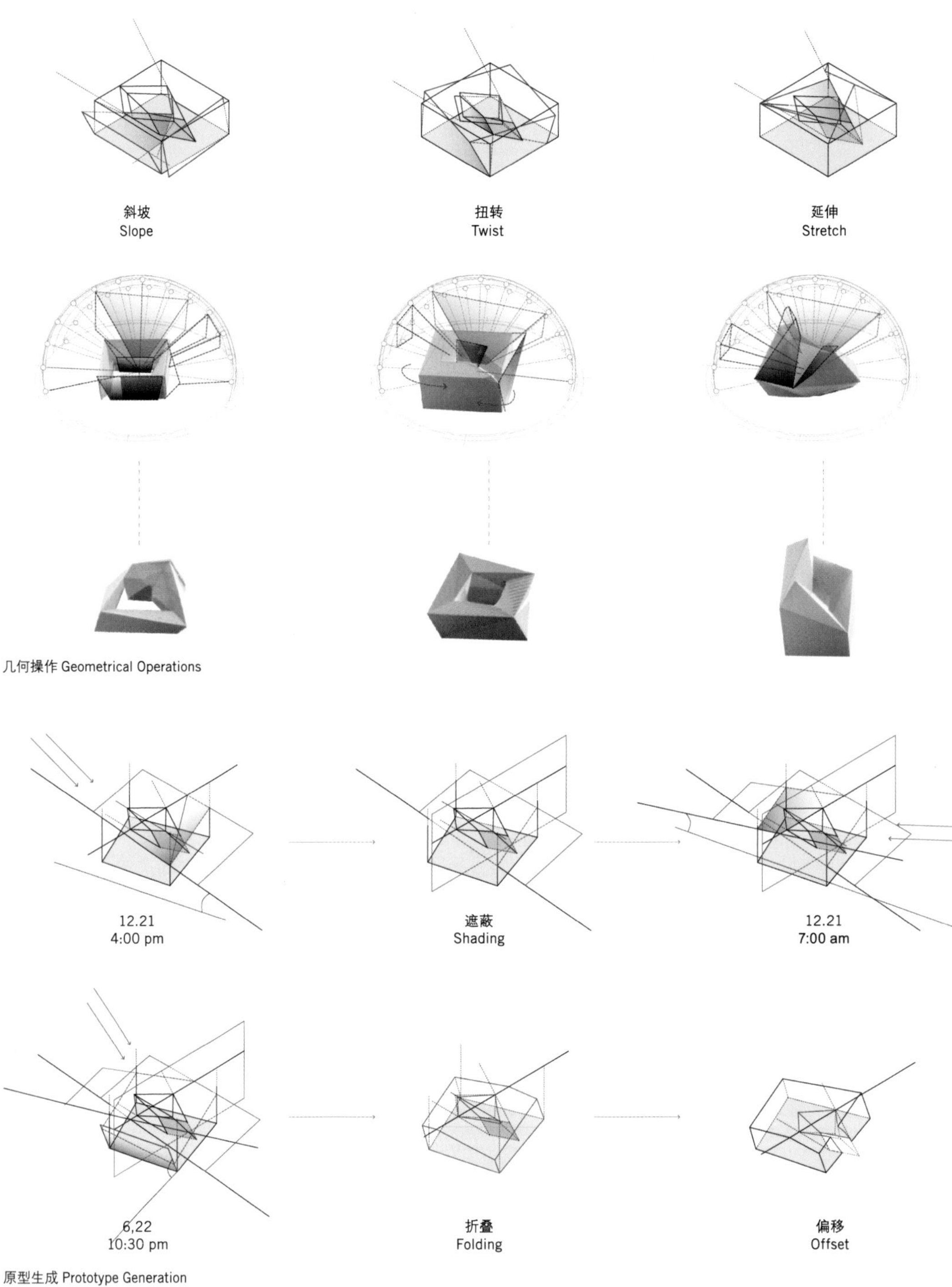


几何操作 Geometrical Operations

原型生成 Prototype Generation

几何单元采光优化
Unit Geometric Lighting Optimization

该原型设计利用几何形状来进行遮阳和采光。核心控制元素是在不同季节的特定太阳高度角，以解决强烈的光线和西向暴光的问题。通过对自然民居的研究，一些原型操作手法可被借鉴、用来优化。

The prototype is designed to make use of geometric shape for shading and lighting. The core control elements are specific solar elevation angles which can solve the strong light and west exposure problems in different seasons. Through the study of the natural settlements, several operations are adopted for optimizing.

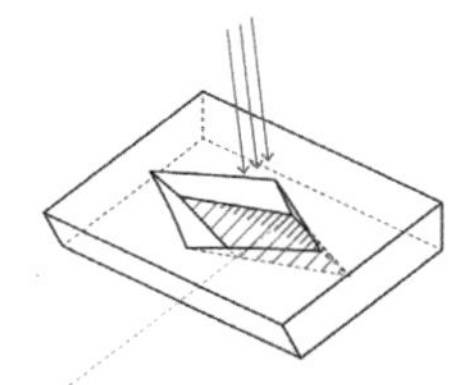

夏天的院子：阴影
Courtyard in Summer: Shadow

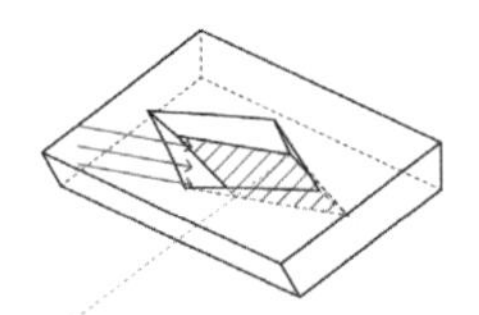

冬天的院子：直射阳光
Courtyard in Winter: Direct Sunlight

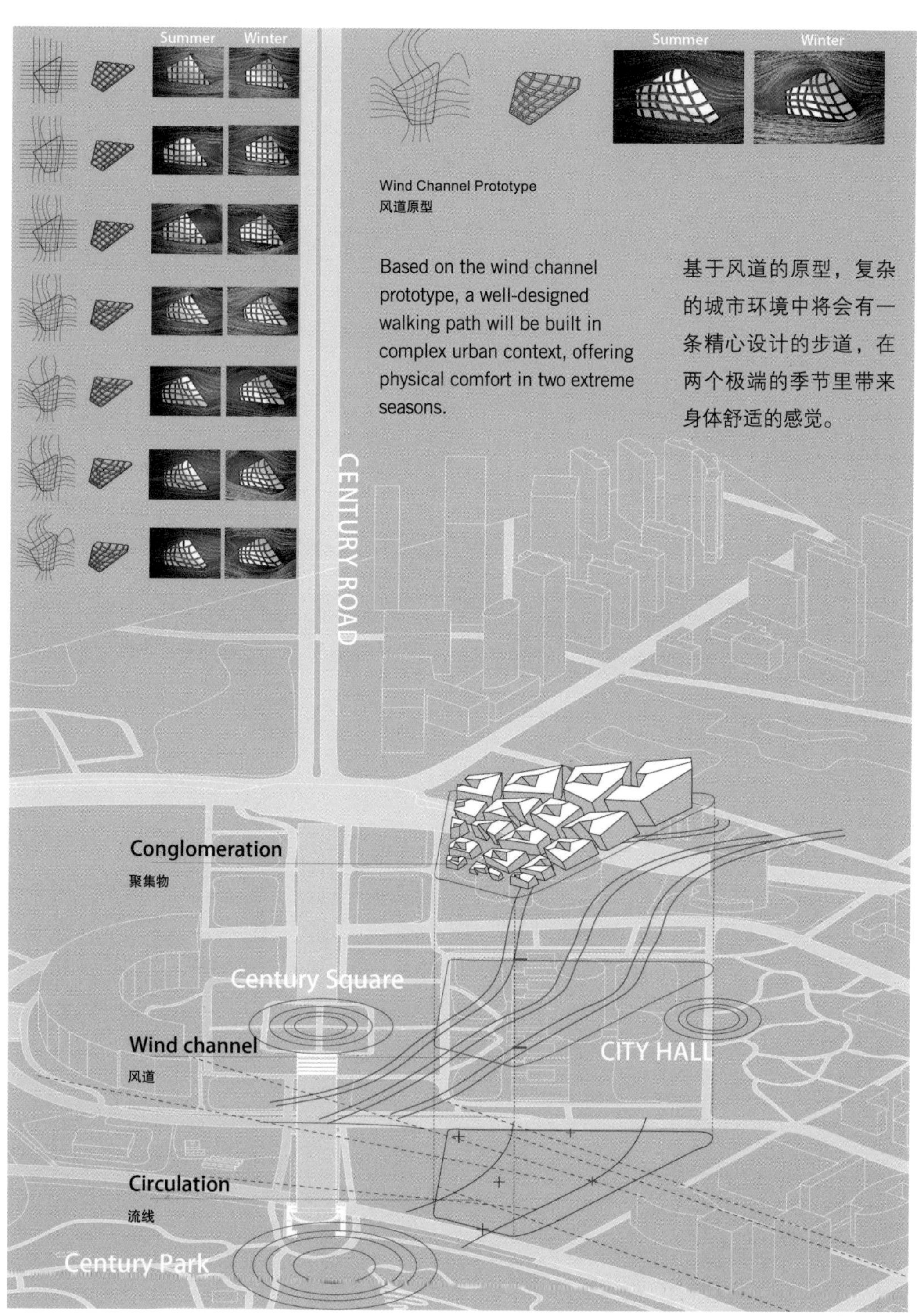
Summer
Winter
Wind Channel Prototype
风道原型
Based on the wind channel prototype, a well-designed walking path will be built in complex urban context, offering physical comfort in two extreme seasons.
基于风道的原型，复杂的城市环境中将会有一条精心设计的步道，在两个极端的季节里带来身体舒适的感觉。
CENTURY ROAD
Conglomeration
聚集物
Century Square
Wind channel
风道
CITY HALL
Circulation
流线
Century Park

布局通风研究
Layout Ventilation Study

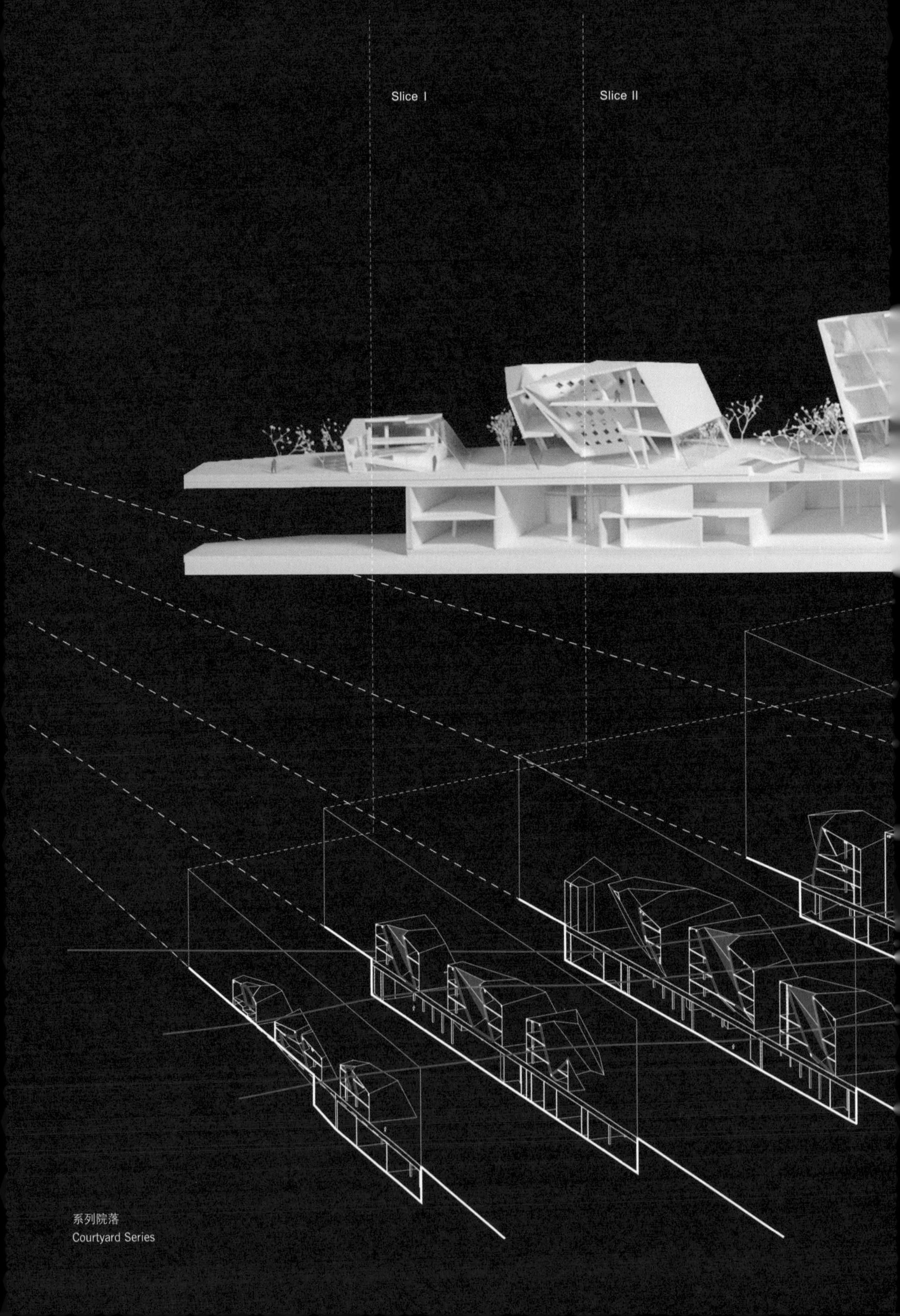


系列院落
Courtyard Series

Slice Ⅳ

Slice Ⅴ

wind corridor

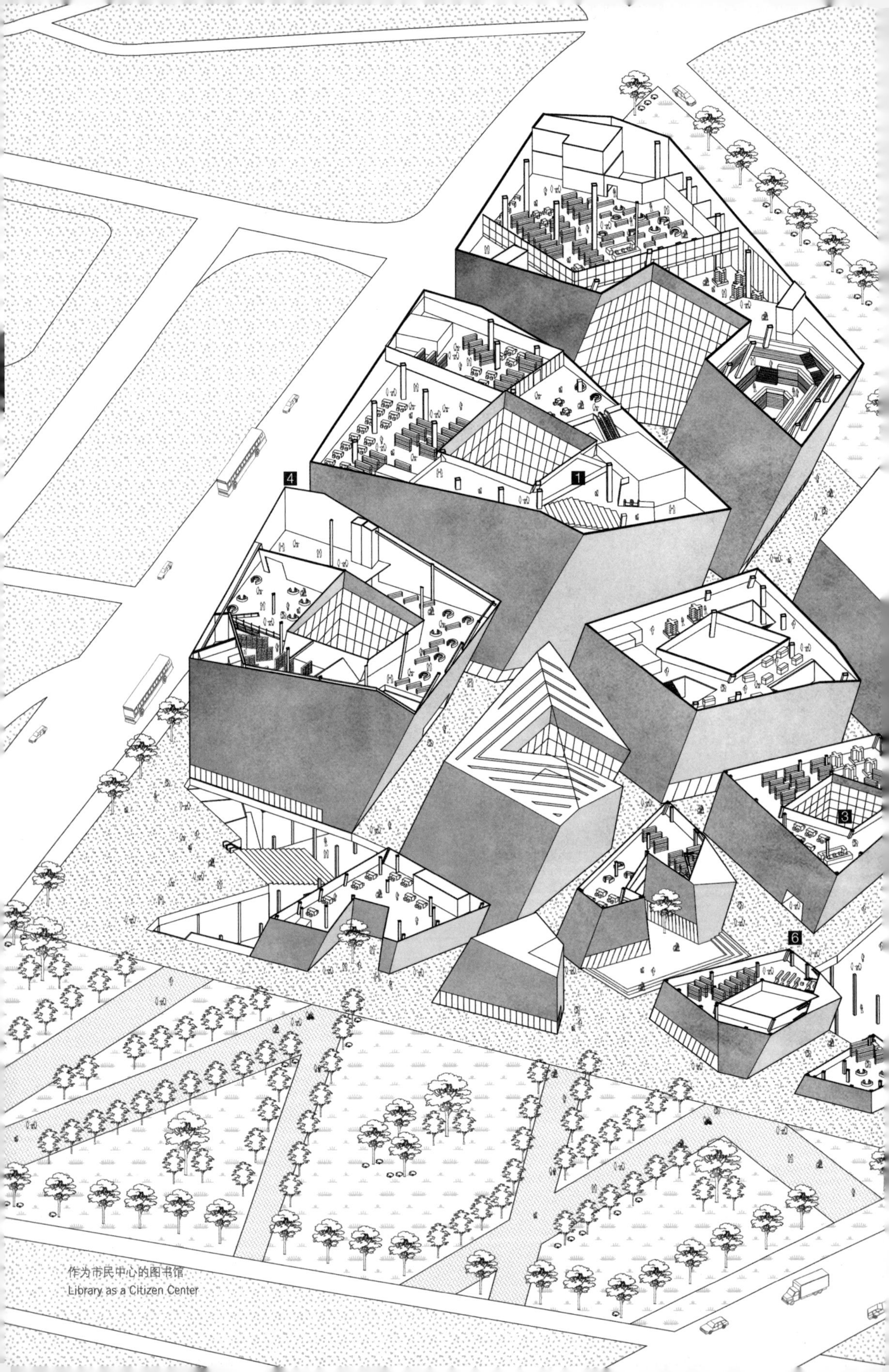

作为市民中心的图书馆
Library as a Citizen Center

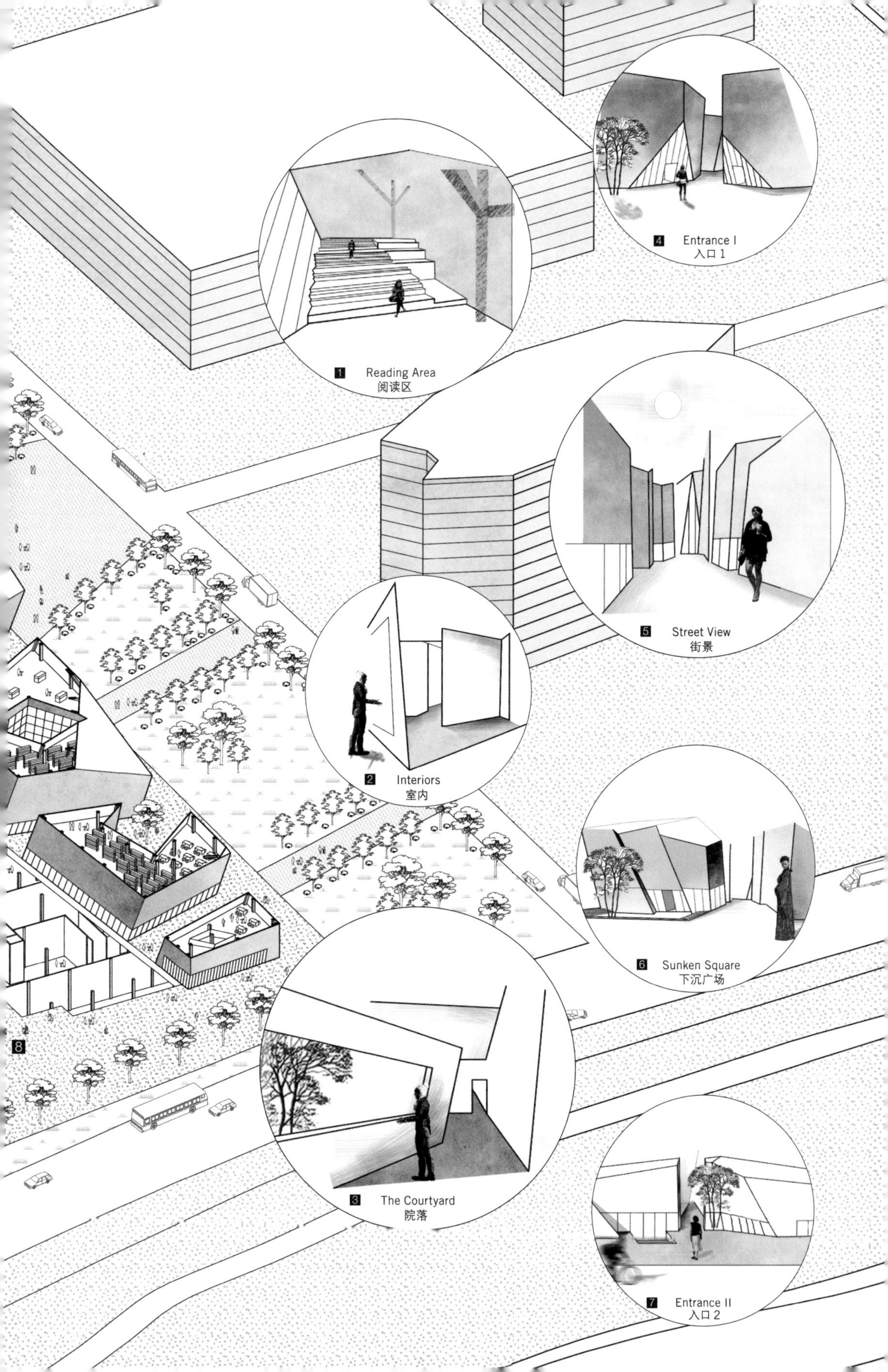

1 Reading Area
阅读区
2 Interiors
室内
3 The Courtyard
院落
4 Entrance I
入口 1
5 Street View
街景
6 Sunken Square
下沉广场
7 Entrance II
入口 2
8

由于上海经常下雨，所以立面排水非常重要。黏土若长时间浸泡在水中，会逐渐被侵蚀和污染。将不同类型的纹理在水下进行测试，以优化排水能力。V形纹理虽然简单，但效果很好。

Since it rains frequently in Shanghai, facade drainage is very important. Clay can be gradually eroded and stained if immersed in water for a long time. Different types of texture are tested under water to optimize the drainability.The V shape texture, though simple, works efficiently.

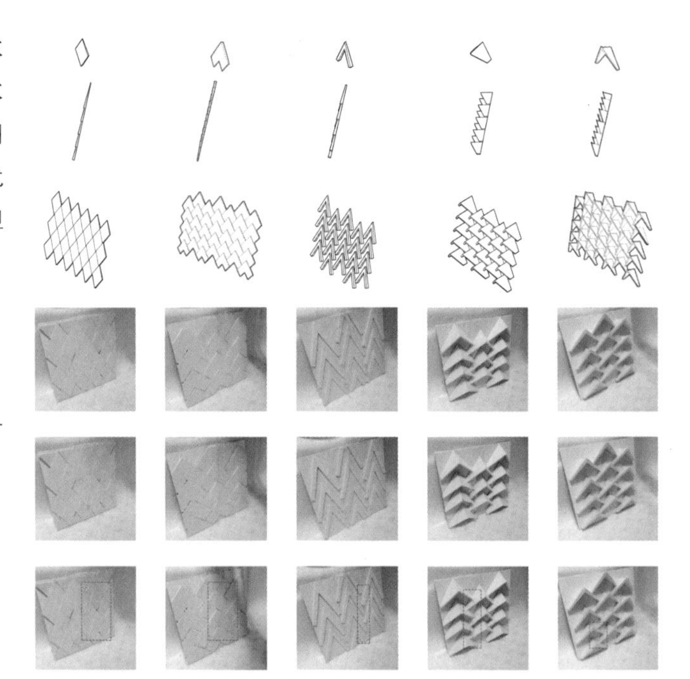

滤水性能测试
Drainability Tests

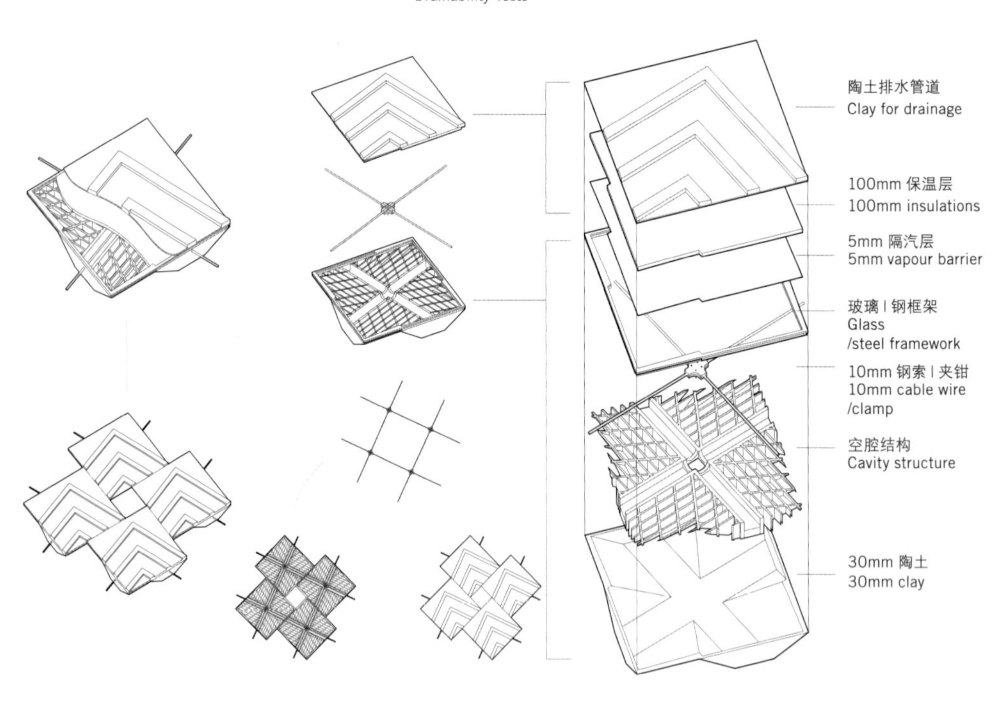


细部研究
Detail Research

9:00　　12:00　　15:00

Summer　夏天

Winter　冬天

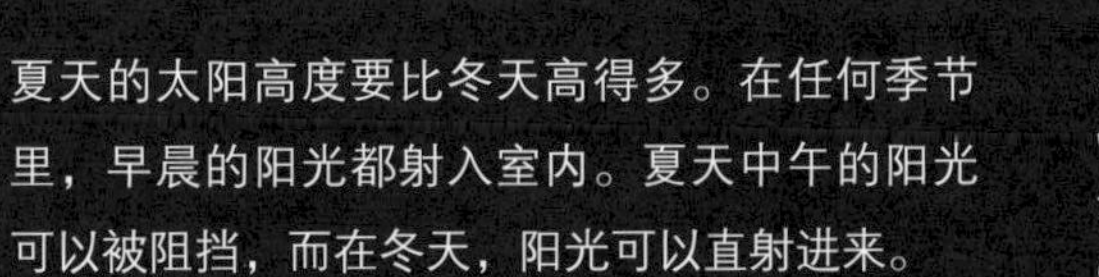

夏天的太阳高度要比冬天高得多。在任何季节里，早晨的阳光都射入室内。夏天中午的阳光可以被阻挡，而在冬天，阳光可以直射进来。

Solar altitude is much higher in summer than it is in winter. In any season, moring sunlight can shine into the room. Sunlight at summer noon can be blocked out while in winter direct sunlight can come in.

# 城市更新中的热塑形

# Heat Formation in Urban Regeneration

基地位于上海市中心城区的中环与内环之间，紧邻陆家嘴金融城、外滩金融集聚带。黄浦江呈 U 形大转弯，形成了杨浦区三面环水的滨江地脉特征。杨浦区滨江岸线长达 15.5km，是上海中心城区中最长的江岸线。杨浦滨江占上海中心城区黄浦江西岸线长度近 1/3，是黄浦江北段发展带的重要发展区域。基地上海杨树浦发电厂位于杨浦区滨江工业区，杨树浦路以南，东临上海国际时尚中心（上海第十七棉纺织总厂），西部多为老工业区，有大量工业遗存。

The base is located in the central city of Shanghai, between the central and inner ring of the central city, close to the Lujiazui Financial City and the Bund financial gathering belt. Yangpu Riverside occupies nearly one-third of the West Bank in length in the central city. It is an important area for the development zone of the northern section of the Huangpu River. The Huangpu River has a U-shaped turn, forming the characteristics of the riverside veins on the three sides of the Yangpu District. The Jiang'an Line is 15.5km long in Yangpu District and is the longest riverbank in Shanghai's downtown area. Our site Shanghai Yangshupu Power Plant is located in Binjiang Industrial Zone, Yangpu District, south of Yangshupu Road, east of Shanghai International Fashion Center (Shanghai No. 17 Cotton Textile General Factory), and the western part is mostly old industrial area with a large number of industrial remains.

指导老师：李麟学、周渐佳
评审老师：薛广庆、李卓
助教：何美婷、吕悠、烘烽桓、葛康宁
学生：张琪、李昊、甘崇雨、周与锋、李安贵、唐倩倩、雒雨、郑国臻
Tutors: LI Linxue, ZHOU Jianjia
Review Teachers: XUE Guangqing, LI Zhuo
Teaching Assistants: HE Meiting, LYU You, HONG Fenghuan, GE Kangning
Students: ZHANG Qi, LI Hao, GAN Chongyu, ZHOU Yufeng, LI Angui, TANG Qianqian, LUO Yu, ZHENG Guozhen

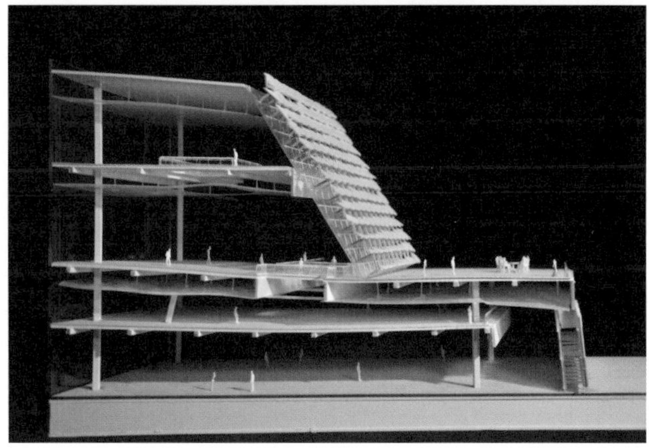

左页：蚁穴模型照片
右页：冷热流模型照片
Left Page: Model photo of Ant Colony
Right Page: Model photo of Cold and Hot

本设计从对自然界蚁穴的空间结构研究入手，将处理建筑中热量传播的重点放在了上海北外滩杨树浦自来水厂改造，即如何利用旧有建筑创造新的功能。此外，为解决不同季节及不同地区主导风向不同的问题，并且为将来使用提供较好的空间，我们设计了不同的通道结构。通道的大小和连接方式取决于于风的利用方式，使得夏季风更多地穿过建筑，冬季风更少地影响建筑内部的使用。为此我们想到在建筑的表皮设计不同孔隙率的材料，从而满足不同的需求。在通道结构形式的选择上，我们借鉴了参数化设计的手段，对比了不同的参数从而得到一个适应环境的较优解。

This design starts from the study of the spatial structure of natural antholes and focuses on the transformation of the heat transfer in the construction of the Yangshupu Waterworks in Shanghai North Bund, and how to use the old buildings to create new functions. In addition, to respond to the change of dominant wind direction in different seasons and different regions, and provide better space for future use, we designed different channel structures. The size and the way of connection of the passageway depend on the use of the wind, so that the summer monsoon can pass through the building more, and the winter wind can affect the interior less. We use different materials in the skin of the building to meet different needs. As for the channel structure, we take parametric design method as the reference and compare different parameters to obtain the best solution to adapting the environment.

# 蚁穴
# Ant Colony

学生 Students：
李昊 LI Hao
甘崇雨 GAN Chongyu

指导老师 Tutors：
李麟学 LI Linxue
周渐佳 ZHOU Jianjia

## 热力学建筑原型 Thermodynamic Architectural Prototype

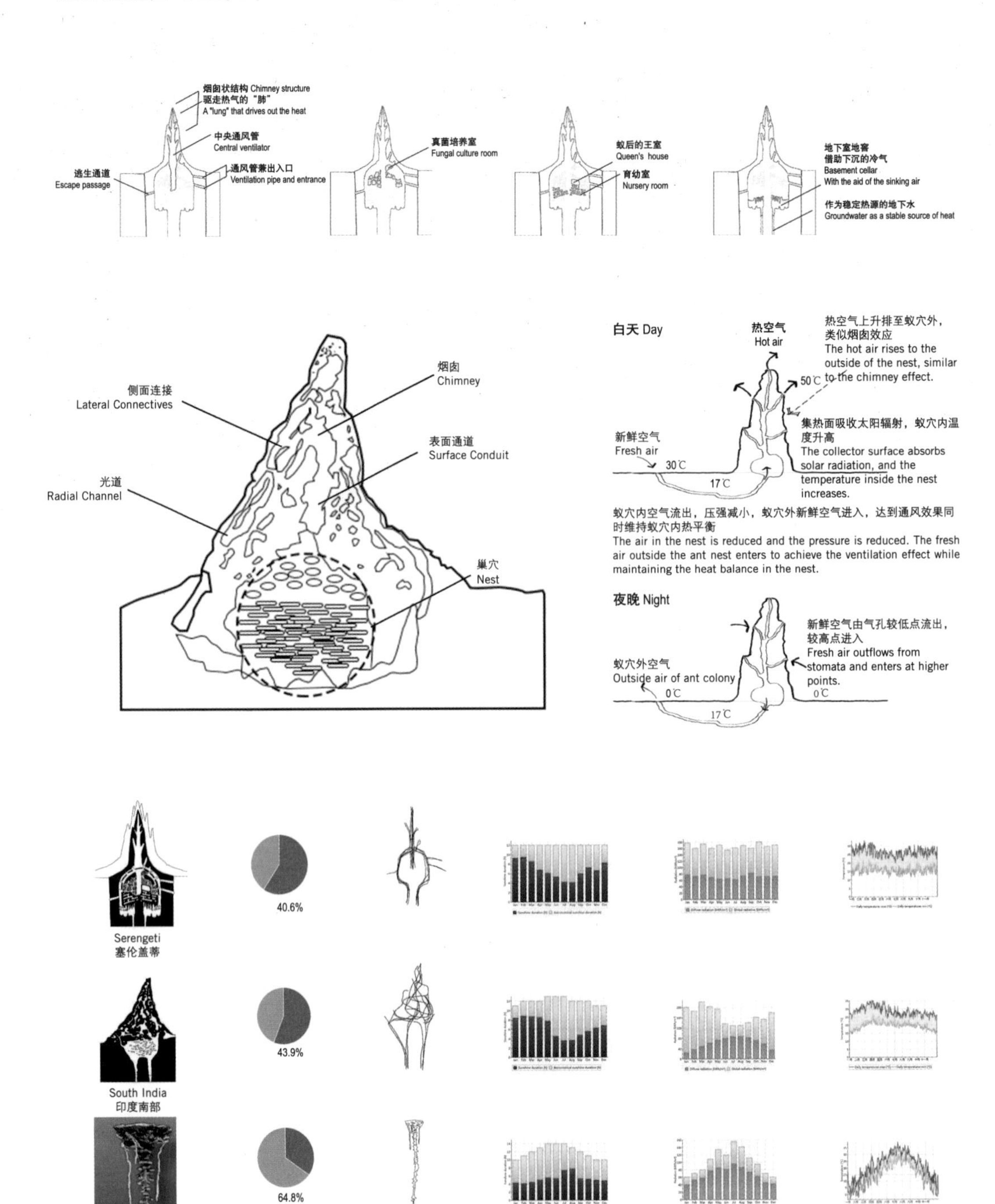


理论研究：蚁穴

Theoretical Research: Ant Nest

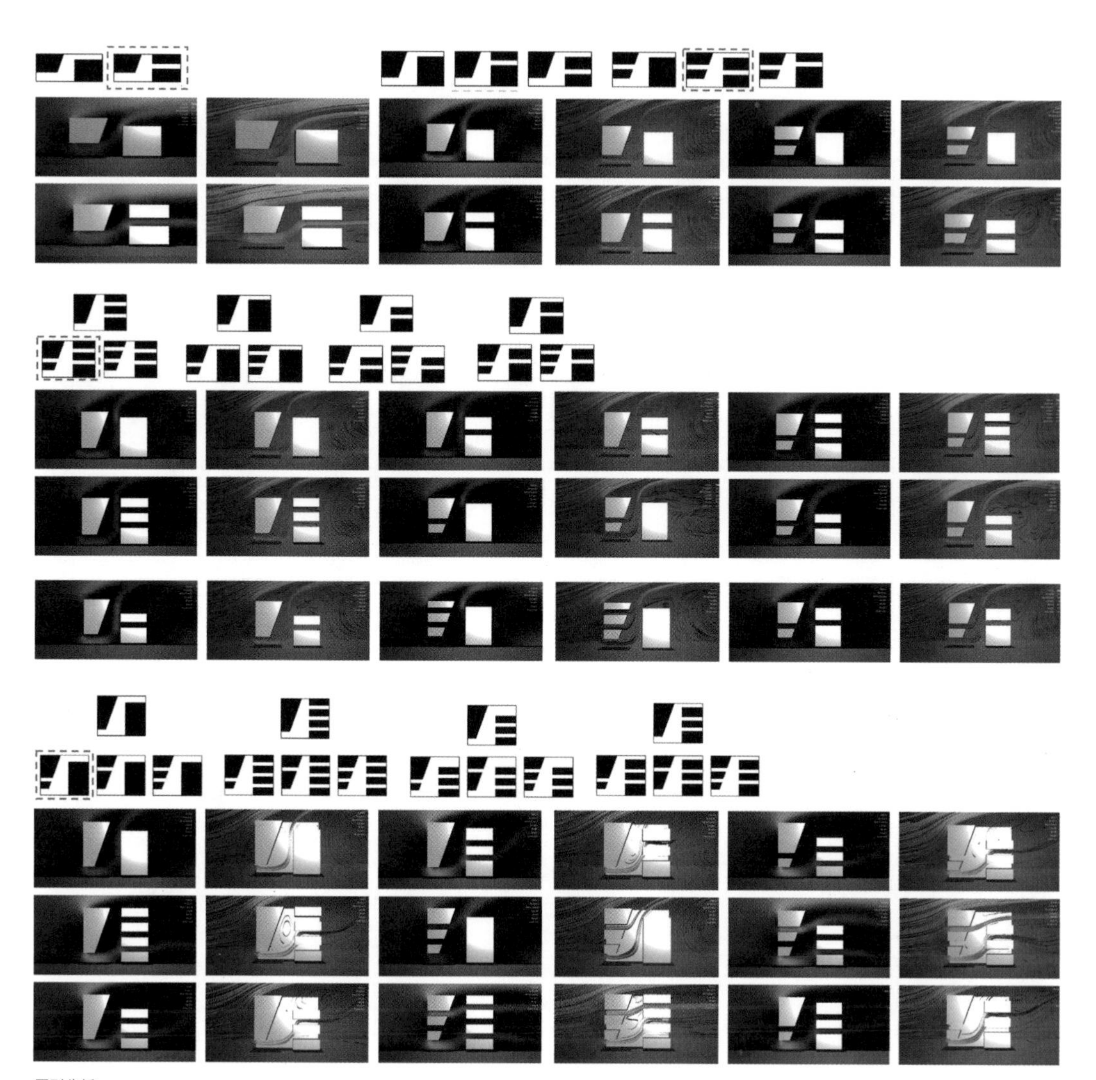

原型分析
Prototype Analysis

提出“蚁穴”概念——结合案例调研的思考，并考虑滨江的气候环境，设计中对水气、温度、物质等调控的需要受到蚁穴的启发。通过对蚁穴的空间结构和热量传播特征分析得出以下结论：维持建构体内部的热平衡需要稳定热源、多孔隙率通道、多重表面等要素。

We propose the concept of “ant colony” because the requirements in regard of moisture, temperature, and mass in our design are all inspired by ant nests. Analyses of the spatial structure and the heat transmission features conclude that heat balance within the construction requires stable heat sources, multiple porosity channels, and multiple surfaces.

## 热力学建筑原型 Thermodynamic Architectural Prototype

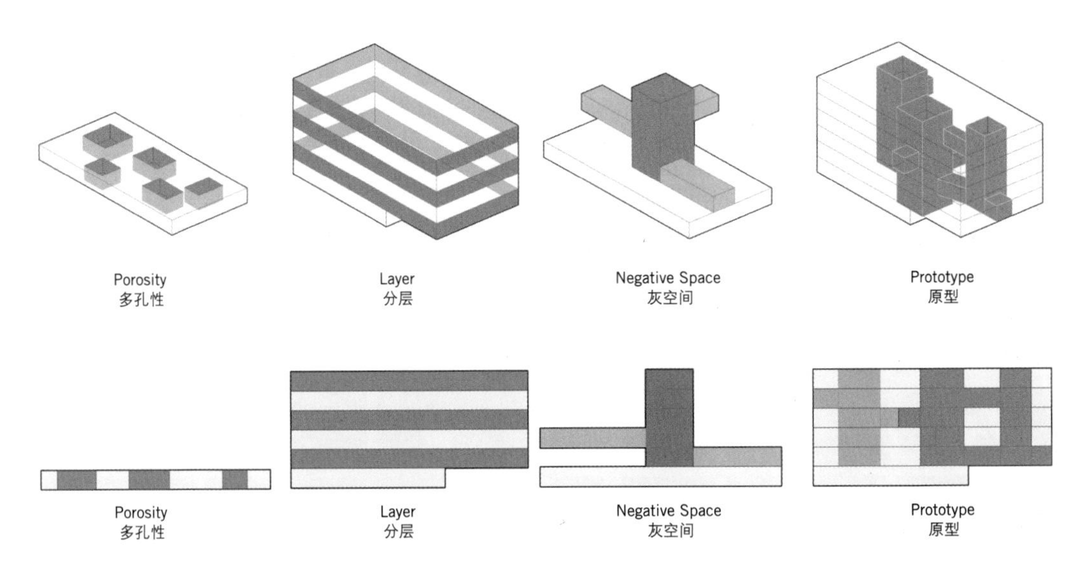


Generation: Community 生成分析：连通性

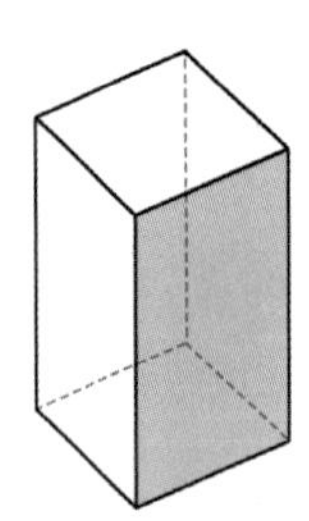

General buildings are enclosed and lighted from south facade
一般建筑朝南面围合并且采光

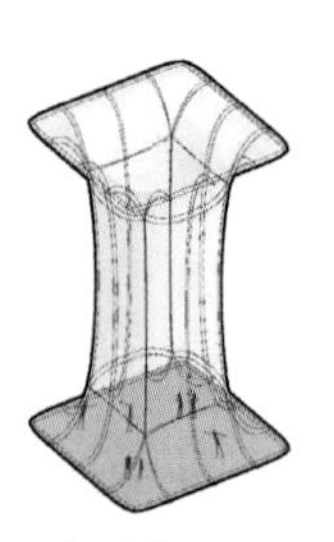

The increase of public space provides more chances for activities
公共空间的增加带来更多的活动

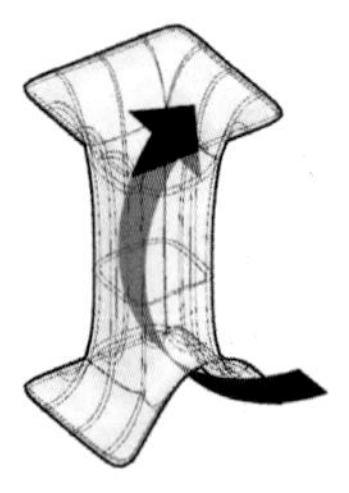

The opening of the courtyard attracts the wind flowing through the building
庭院的开启为建筑引入通风

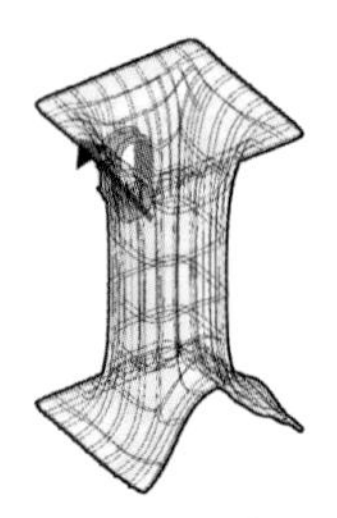

Longitudinal tunnel creates the way of wind flowing
垂直向通道产生风流

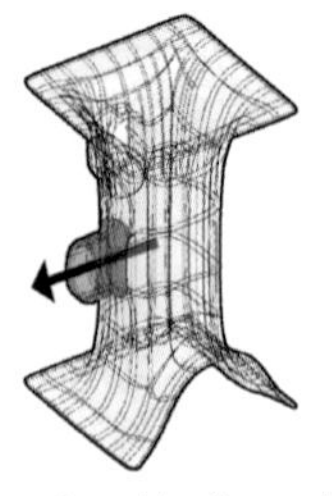

Horizontal tunnel provides the corridors for transportation
水平向通道提供交通组织的过道

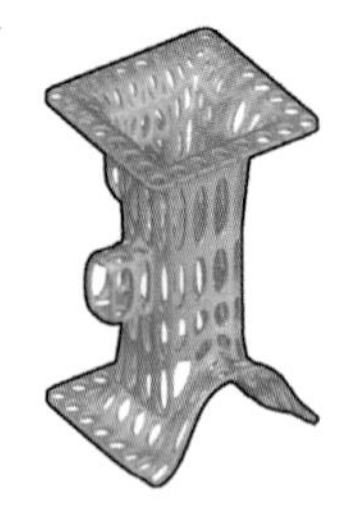

Perforated facade offers more possibilities for daylighting and ventilation
穿孔立面为采光和通风提供了更多的可能性

Generation: monomer 生成分析：单体

生成分析
Generation Analysis

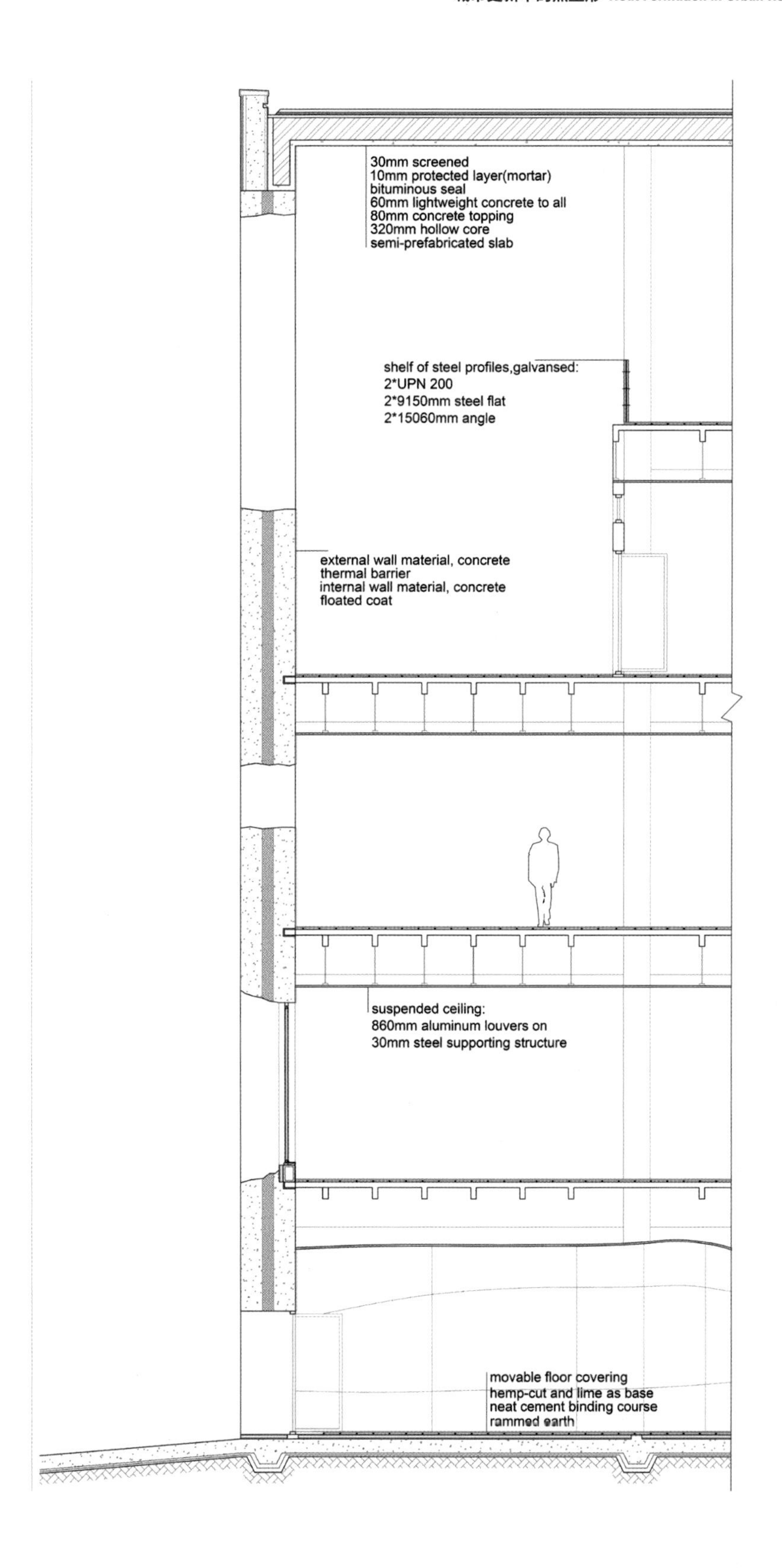


墙体细部
Wall Detail

基地位于上海杨树浦发电厂沿黄浦江一带，基地内保留两个105米高的烟囱和原锅炉房主体结构。烟囱的锅炉房主体框架作为这片工业区历史的一部分被保留下来。同时，作为连接黄浦江与城市的构筑物，随之而来的是如何处理好新旧关系的问题，以及如何转变发展方式、改进工业对城市的影响。新建筑通过研究能量热塑形和在城市气候环境中的热力学原型，实现城市的更新。

设计策略有以下三点：充分利用烟囱；通过能量研究，形成热力学的博物馆原型；在城市尺度和环境下植入原型。

The base is located at Yangshupu Power Station along the Huangpu River in Shanghai. Two 105-meter-tall chimneys and the main structure of the original boiler room are kept in the base. The main frame of the chimney boiler house is preserved as a historical part of the industrial area. At the same time, as the connection of the Huangpu River and the city, it follows the problem of how to deal with the new and the old, and how to change the way of development and improve the impact of industry on the city. The new building realizes urban renewal by studying the thermal prototyping and the thermodynamic prototype in the urban climate environment.

Design strategies include: make full use of the chimney; through energy research, form the museum prototype of thermodynamics; implant prototype in urban scale and environment.

# 冷热流

# Cold and Hot

学生 Students：
唐倩倩 TANG Qianqian
李安贵 LI Angui
郑国臻 ZHENG Guozhen

指导老师 Tutors：
李麟学 LI Linxue
周渐佳 ZHOU Jianjia

## 热力学建筑原型 Thermodynamic Architectural Prototype

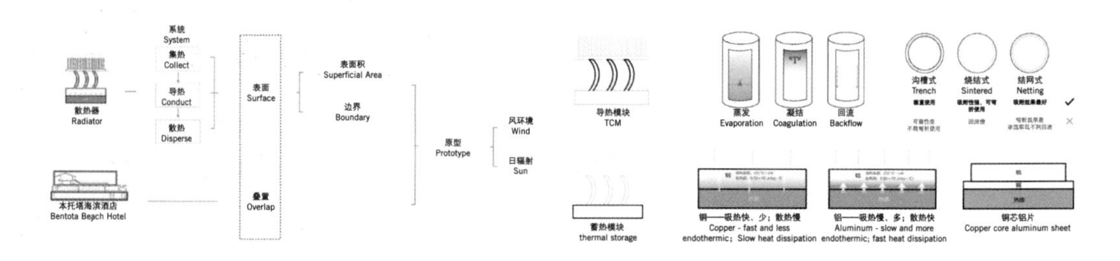


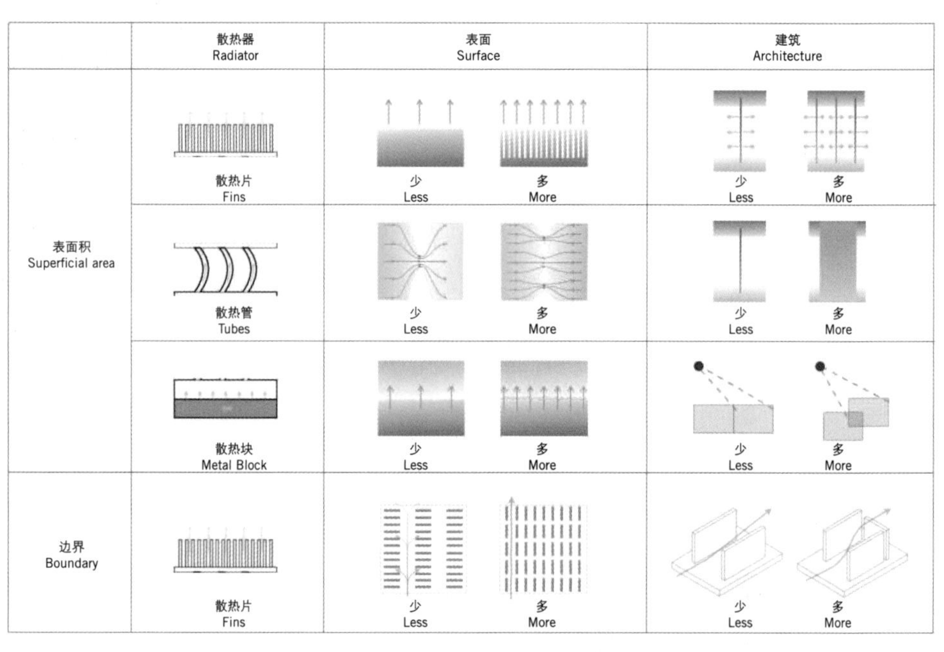


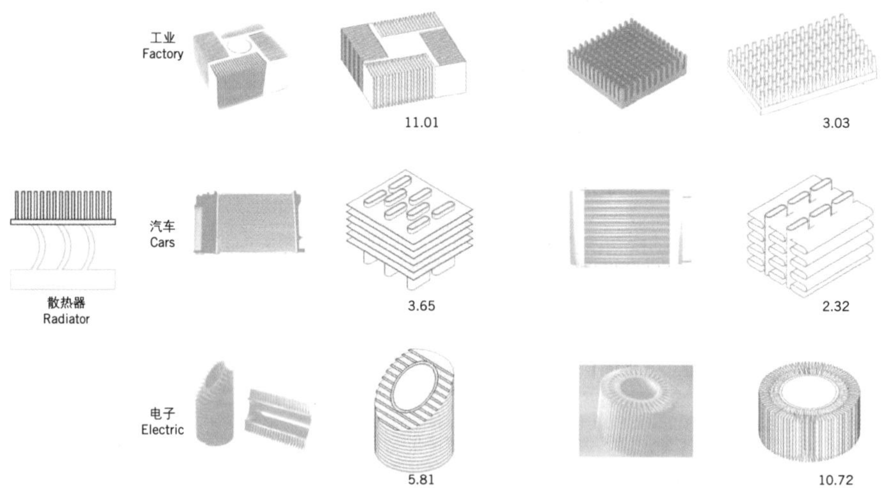


理论研究：散热器
Theoretical Research: Radiator

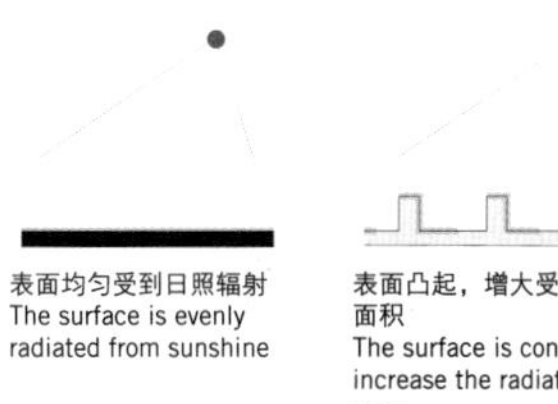
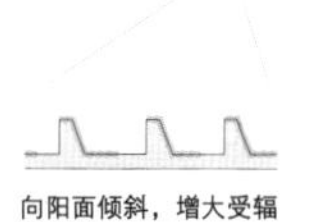

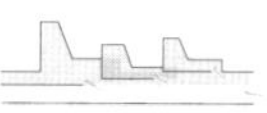

表面均匀受到日照辐射
The surface is evenly radiated from sunshine

表面凸起，增大受辐射面积
The surface is convex to increase the radiation area

向阳面倾斜，增大受辐射面积
The body is inclined to the sun to increase the radiation area.

互相错叠，形成不同辐射程度的空间
Each other is stacked to form a space of different radiation levels.

底层架空，形成进风口
The bottom of the block is built up to form as many inlets as possible.

顶部开口，促进通风
The building opens at the top to promote ventilation

生成设计 Generation

中庭 Atrium

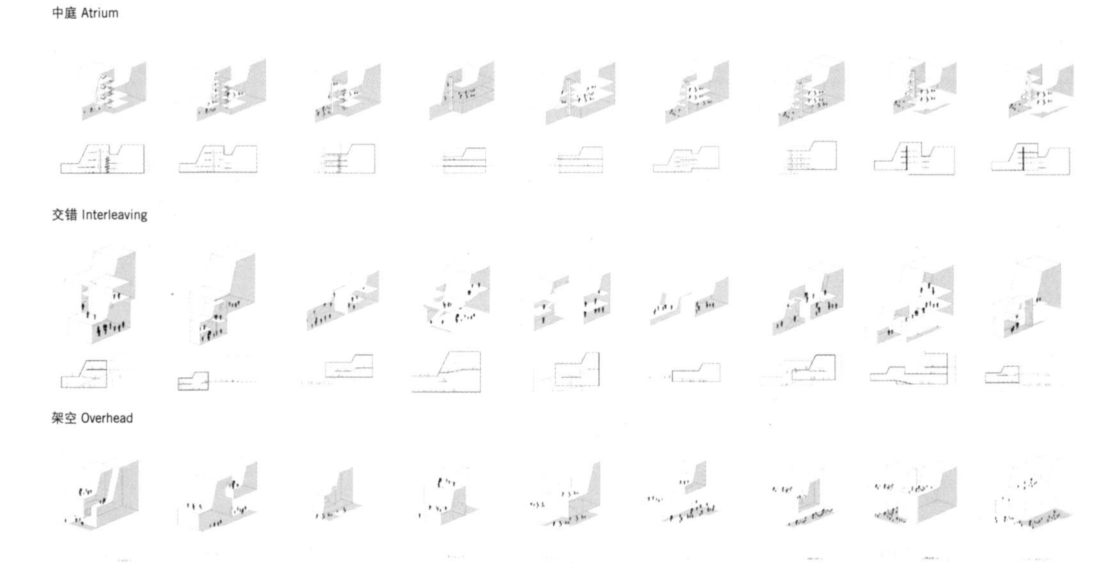

交错 Interleaving

架空 Overhead

空间 Space

原型设计
Prototype Design

基地夏季炎热，从东南方向吹来的夏季风经过黄浦江的冷却，吹入基地内的建筑中。通过烟囱的拔风效应，可以较好地组织基地周边的微气候。项目分析了基地气候、黄浦江水和烟囱等场地特点。挖掘场地的特殊性，将会是项目的优势与亮点。例如，上海夏天炎热，吹东南风，基地的东南方向恰好是黄埔江面。所以当夏季风吹过河面时，由于水气的蒸发升腾，风的温度会下降几度。利用这种自然特性可以为场地甚至城市在夏季带来一丝清凉。

Hot summer wind from the southeast would be cooled down by the Huangpu River before the blow into the buildings on site. Through the wind effect of the chimney, a microclimate around the base can be well organized. The project analyzed the characteristics of base climate, Huangpu River water and chimneys. When the summer winds blow across the Huangpu River to the southeast, the hot temperature of the winds will drop several degrees due to the evaporation of water vapor. This design uses this natural feature to bring a little coolness to the site and even the city in summer.

## 热力学建筑原型 Thermodynamic Architectural Prototype

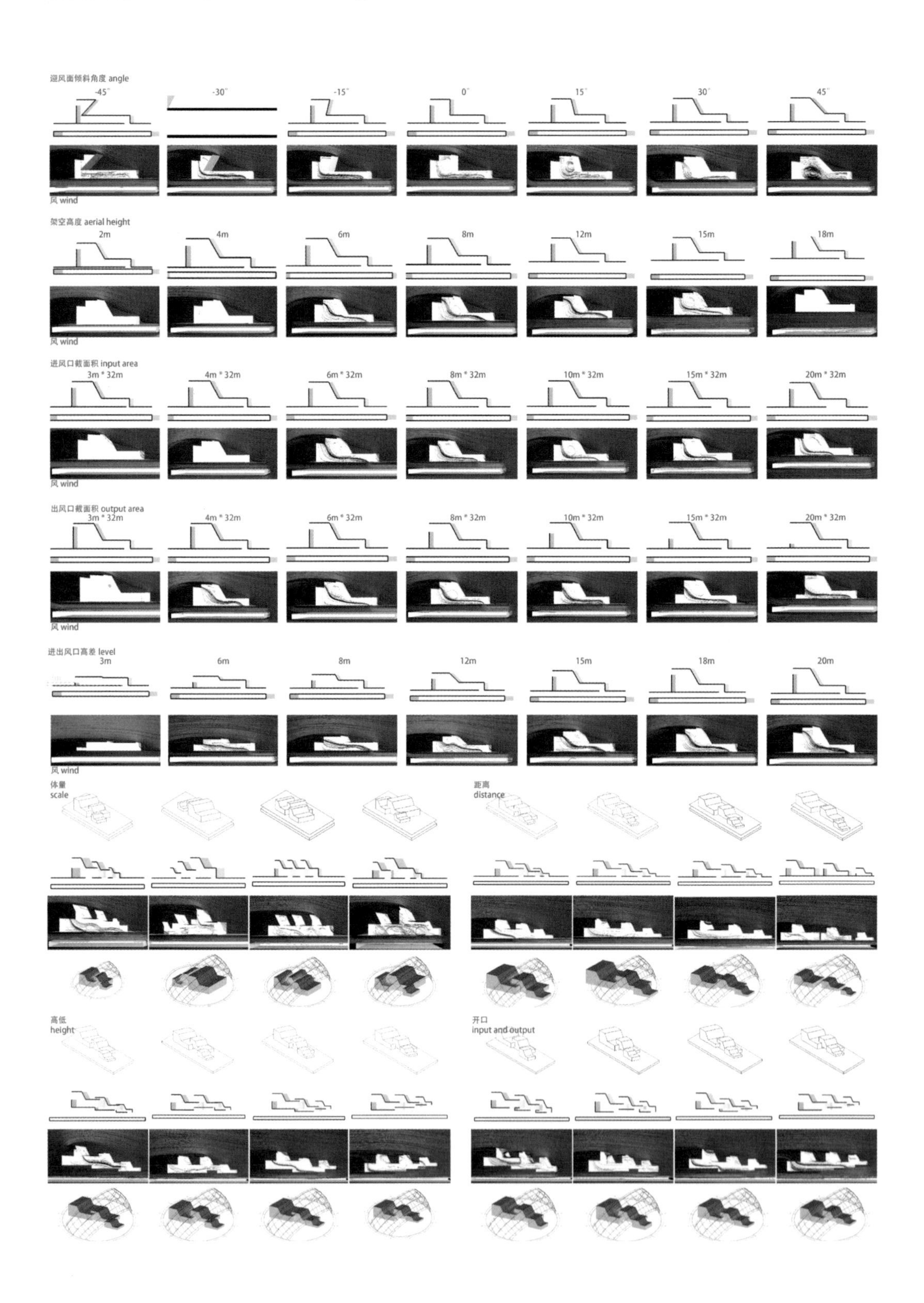

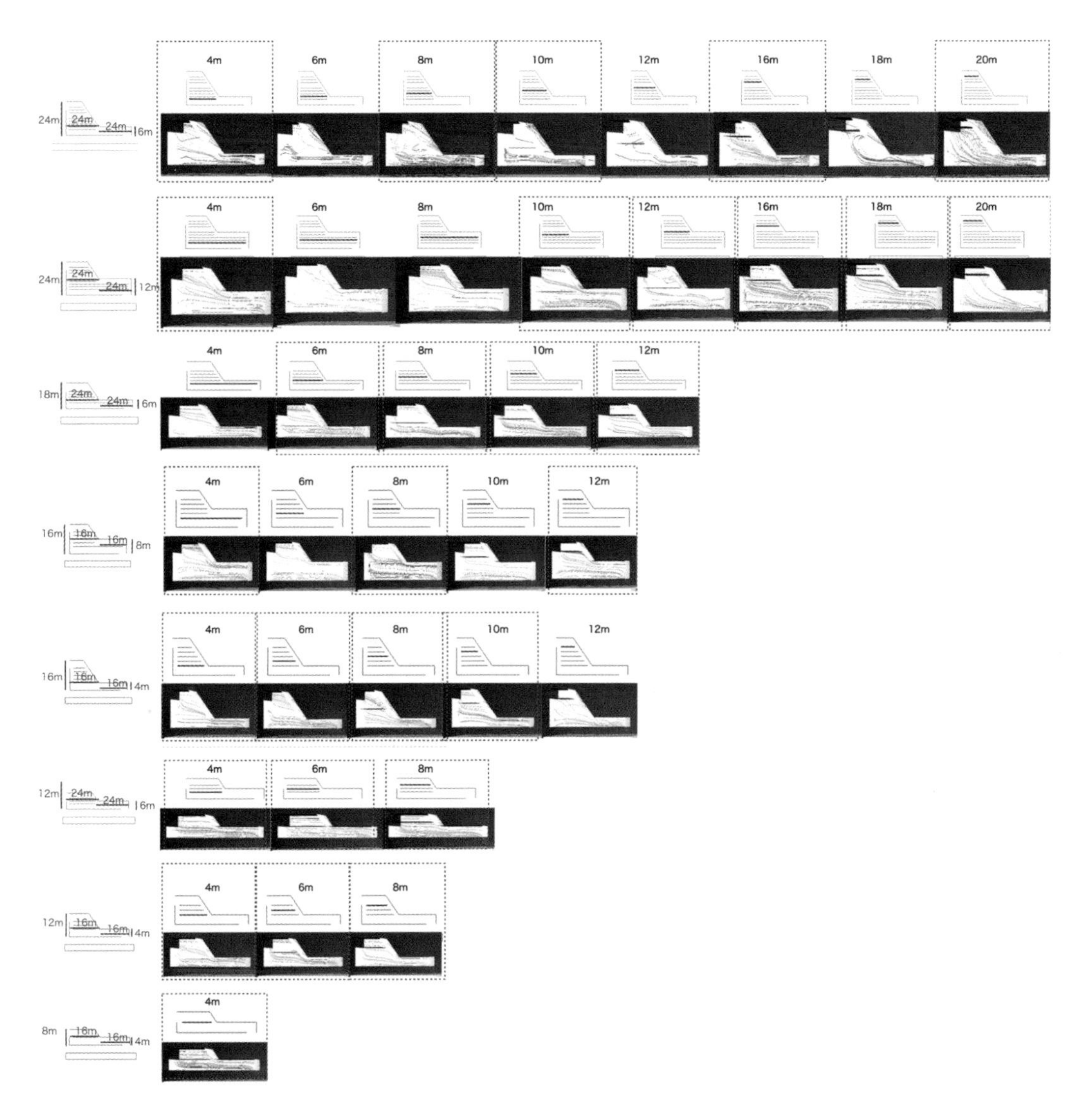


单体楼板通风 Prototype floor ventilation

原型设计
Prototype Design

项目采用散热器的原型来回应这些特性与问题。保留烟囱可以用来维持烟囱效应的产生，促进城市环境通风效果。分别选取宽度、高度和角度作为变量对原型进行通风和日照模拟，以此来优化设计。

The project adopts the prototype of radiator to respond to these characteristics and problems. Preservation of chimneys can be used to maintain chimney effect and promote urban ventilation. We optimized the design by selecting the width, height, and angle as variables to perform ventilation simulation and solar simulation on the prototype.

**热力学建筑原型 Thermodynamic Architectural Prototype**

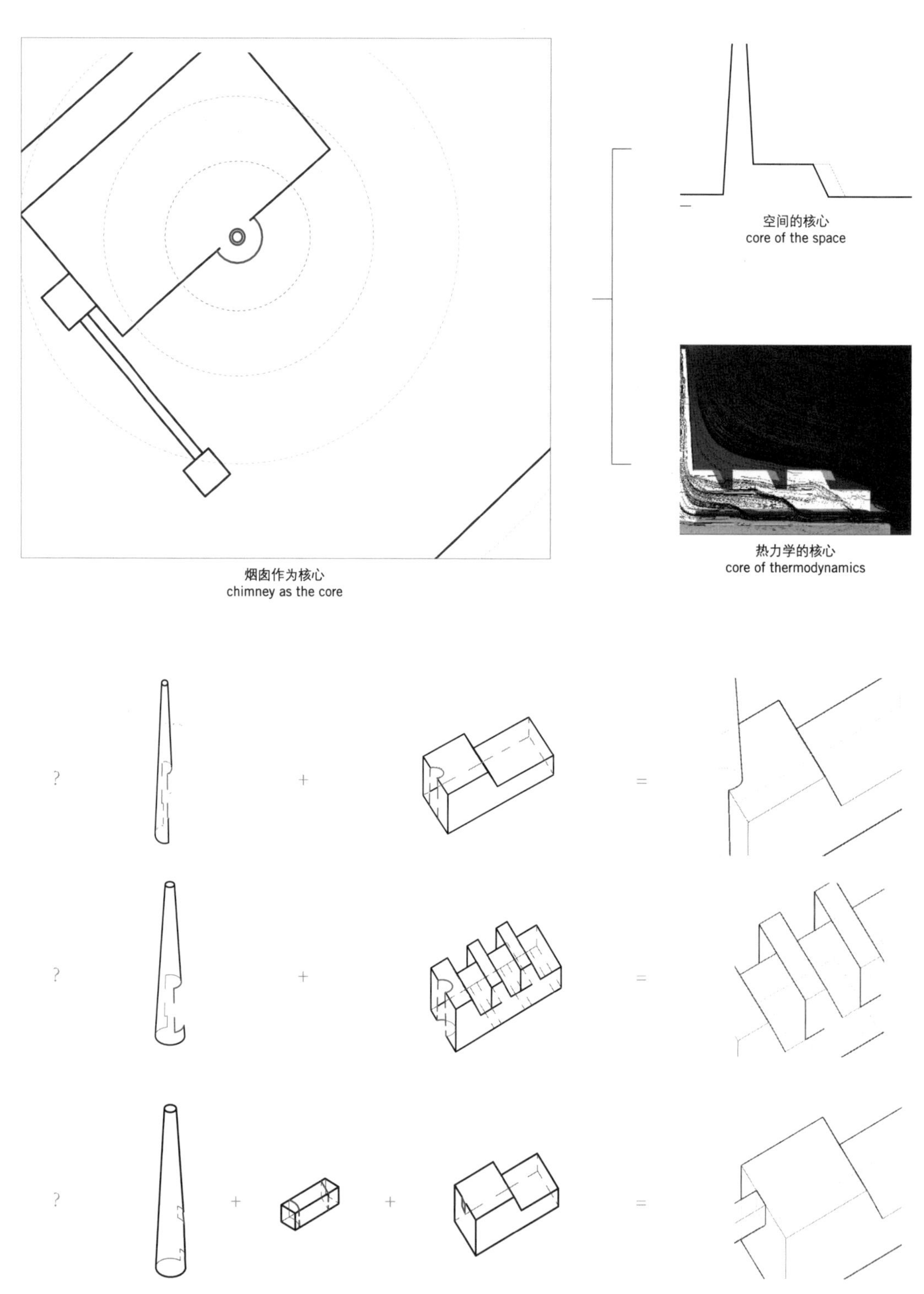


空间的核心
core of the space

热力学的核心
core of thermodynamics

烟囱作为核心
chimney as the core

植入基地
Implantation

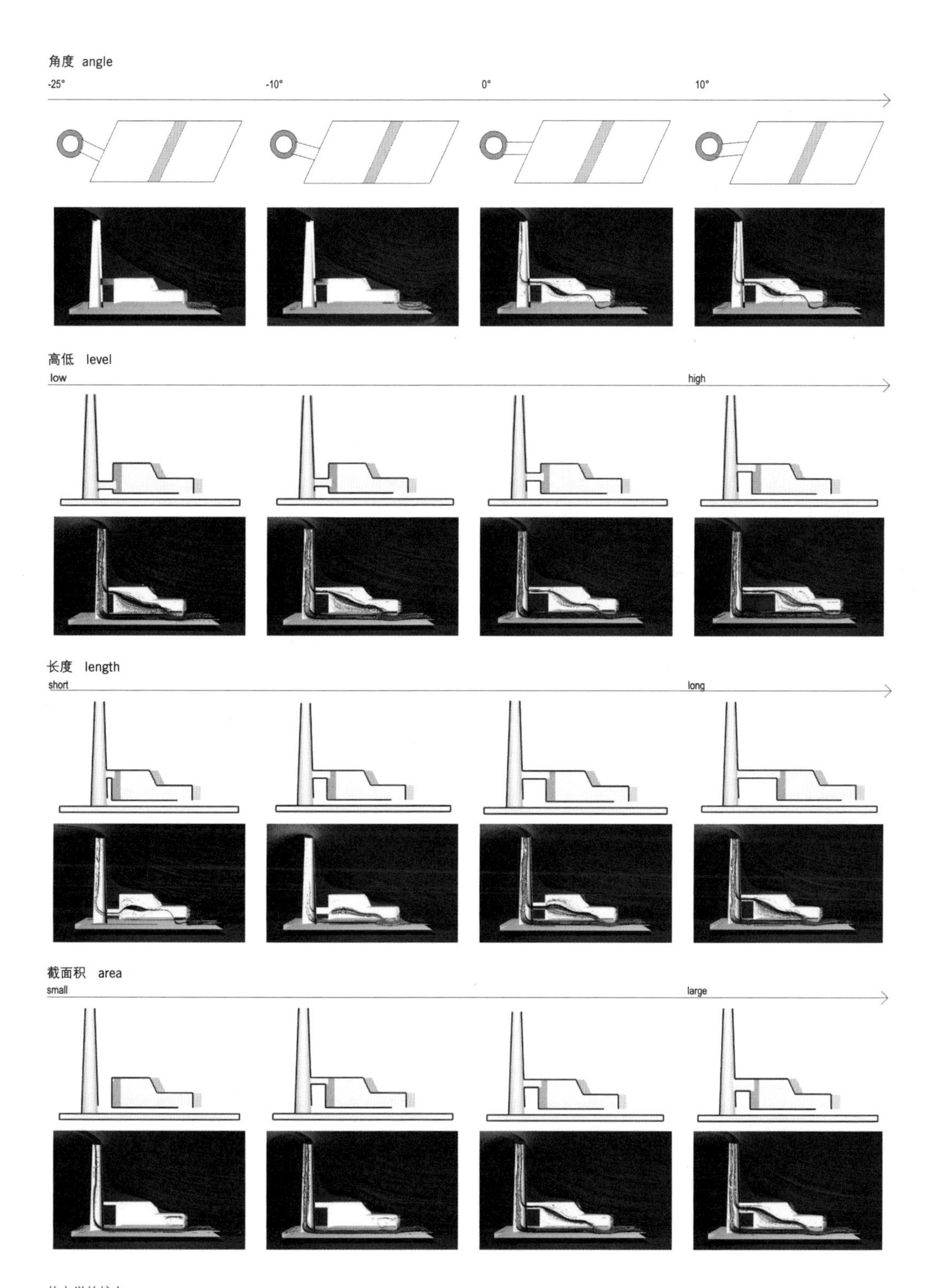


热力学的核心
Core of Thermodynamics

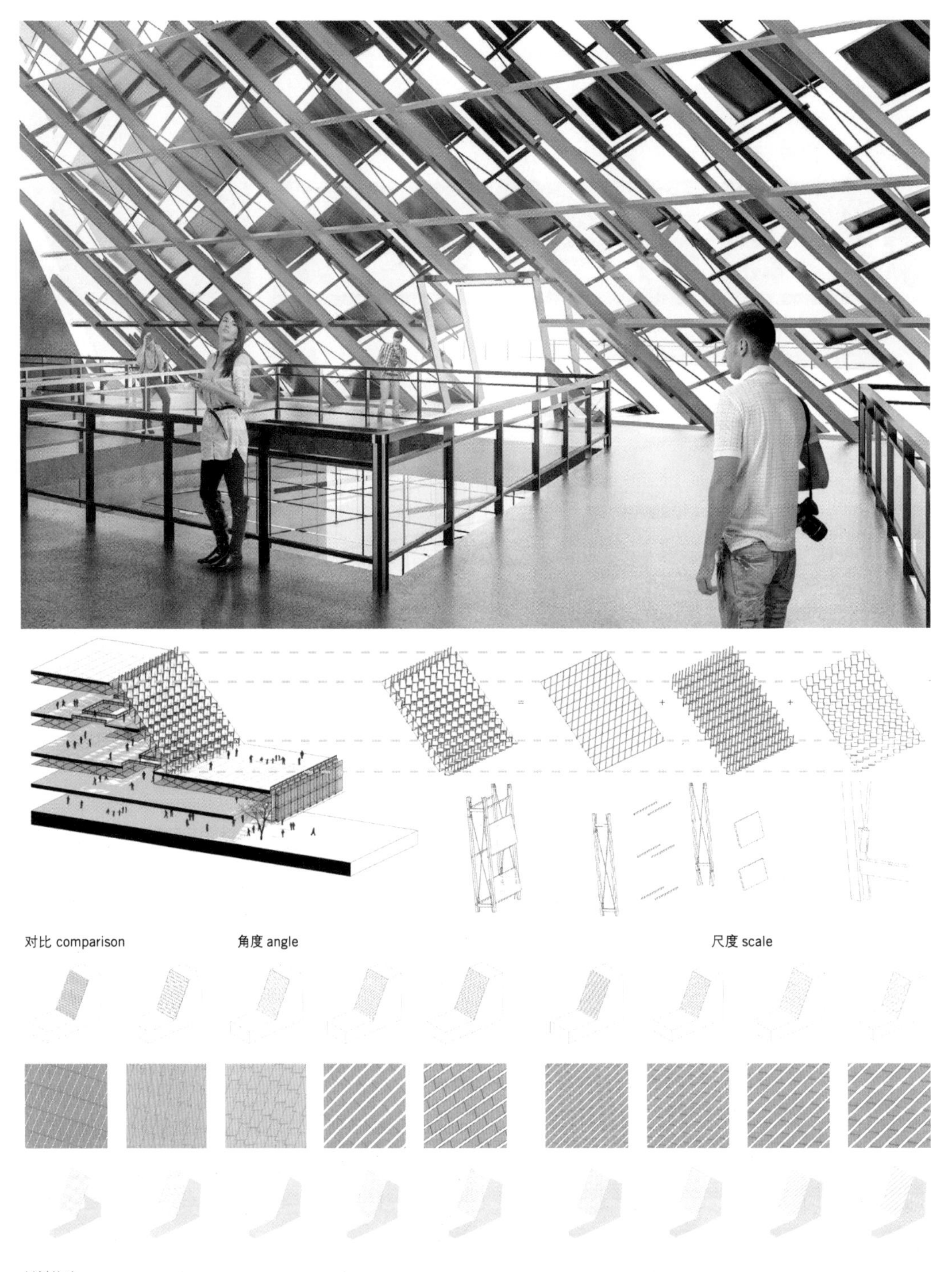


材料构造
Material Structure

热力学模拟

Thermodynamic Simulation

# 自然系统
# Natural System

拟建自然博物馆群，其中本次设计的为一期的自然博物馆，总建筑面积9万平方米，占地24万平方米。项目地址为山东省会济南，在章丘区绣源河畔，是当地重要的历史文化遗产地和自然风光带，自然博物馆群将与周边的文化建筑集群、森林公园、植物园、乡村遗产地等构成重要的城市未来东部文化中心。遗址公园需结合在一期建筑设计之中。

The first phase of the proposed natural museum group has a total construction area of 90,000 square meters and covers an area of 24 hectares. The project is in Jinan, the capital of Shandong Province. It is an important historical and cultural heritage site and natural scenery belt in the area of the embroidered river in Zhangqiu District. The natural museum group will be important to the surrounding culture and cultural clusters, forest parks, botanical gardens and rural heritage sites. The city will be the eastern culture center in the future. The ruins park needs to be integrated in the first phase of the architectural design.

指导老师：李麟学
评审老师：伊纳吉 · 阿巴罗斯 , Renata Sentikiawicz, Max Kuo,
Placido Gonzalez, 周渐佳
助教：何美婷、侯苗苗、苏家慧、郭绵沅津
学生：黄景溢、莫然、陈昌杰、王芮、刘旭田、杨雁容、刘帆、胡雨、薛钰瑾、
布拉德利 · 埃勒布雷克特、康纳 · 斯图西克、克里斯托夫 · 芬克
Tutor: LI Linxue
Review Teachers: Iñaki Ábalos, Renata Sentikiawicz, Max Kuo,
Placido Gonzalez, ZHOU Jianjia
Teaching Assistants: HE Meiting, HOU Miaomiao, SU Jiahui,
GUO Mianyuanjin
Students: HUANG Jingyi, MO Ran, CHEN Changjie, WANG Rui, LIU Xutian,
YANG Yanrong, LIU Fan, HU Yu, XUE Yujin，Bradley Ellebracht,
Conor Stosiek，Kristoff Fink

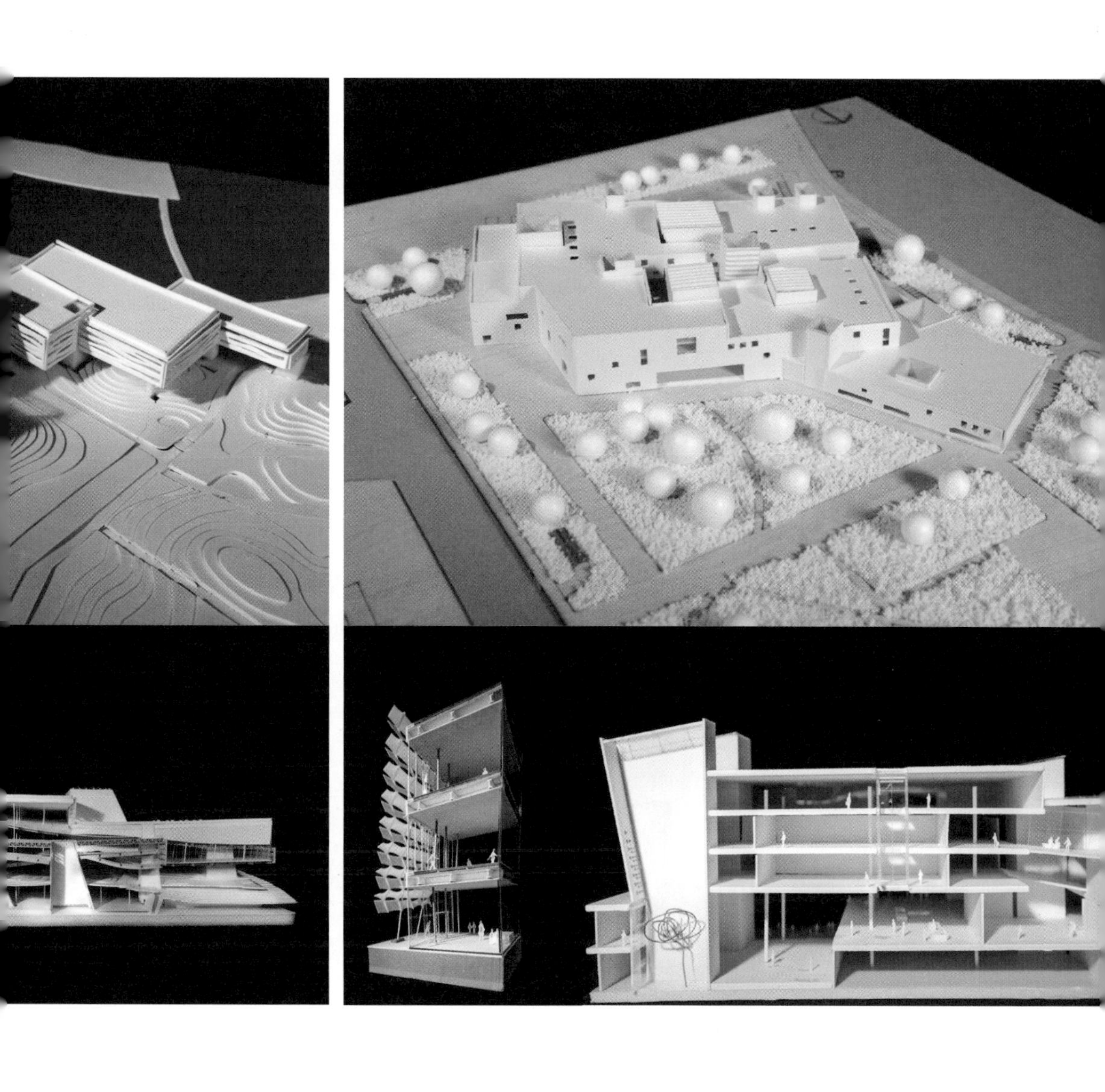

左图：森林模型照片
中图：双系统空气流动模型照片
右图：空腔·孔隙·温室模型照片
Left: Model photo of Forest
Middle: Model photo of Bi-Sourse Reactivity
Right: Model photo of Cavity, Porosity, Green House

通过对自然界现存系统的学习——如森林生态系统、村庄演替、生物呼吸系统等回归最本质的自组织特性，望能从自组织中学习形式追随能量的法则，使建筑具备应对环境变化的调节能力。设计研究流程主要包括从自组织关键词、森林原型的研究，到建筑案例研究，再到具体设计逐步落地；方案推进通过以风、光、热三个自然条件为法则对建筑形态不断进行重塑和检测，从而达到场地上能量流动与建筑功能结合的较优解。

Through the study of the existing systems in nature, such as forest ecosystems, village succession, biological respiratory system, etc., we hope to learn how forms can follow the energy under the control of self-organization to enable buildings adapt to the changing environment. The design research process mainly includes keywords of self-organization, research of forest prototype, architectural case studies and concrete design. The program promotes the continuous remodeling and detection of architectural forms through the three natural conditions-- wind, sunlight and heat to achieve a better solution combining energy design and building functions.

# 森林
# Forest

Student 学生：
LIU Xutian 刘旭田
WANG Rui 王芮
YANG Yanrong 杨雁容
Conor Stosiek 来纳·斯图西克

Tutor 指导老师：
LI Linxue 李麟学

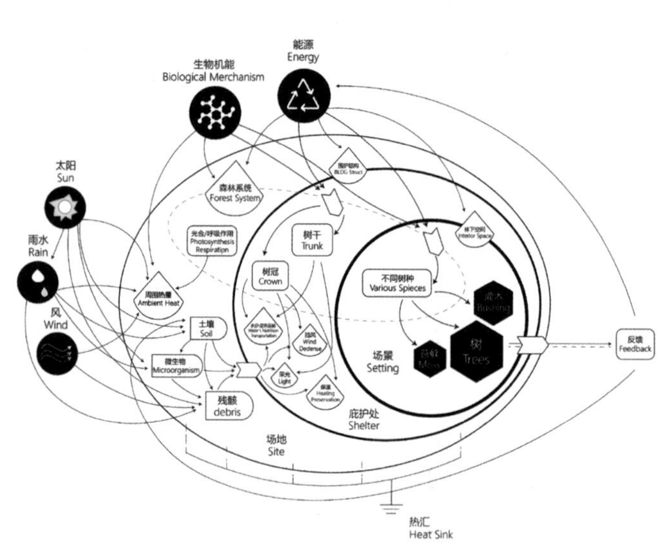


自组织所涵盖的热力学范畴：系统通过与外界交换物质、能量和信息而不断降低自身熵含量，提高其有序度；自组织在热力学范畴内又细分为协同学、耗散结构以及熵减少等。

Self-organization covers the scope of thermodynamics: the system continuously reduces its entropy content and improves its order by exchanging substances, energy and information with the outside world.

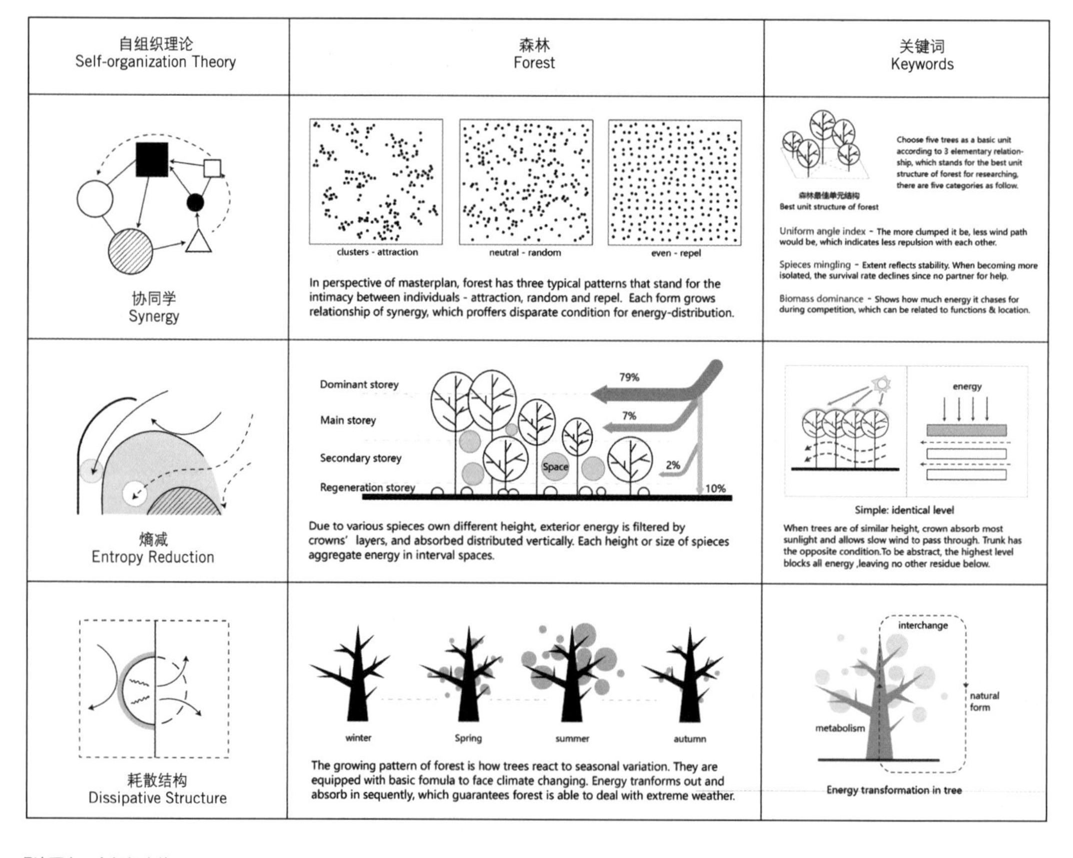


理论研究：自组织森林

Theoretical Research: Forest of Self-organization

Hierarchy 层级

Energy Level

Scale / Contain / Overlap / Sequence / Competence

Aggregation 聚集

Density / Number +Density / Pattern

Adaptation 适应

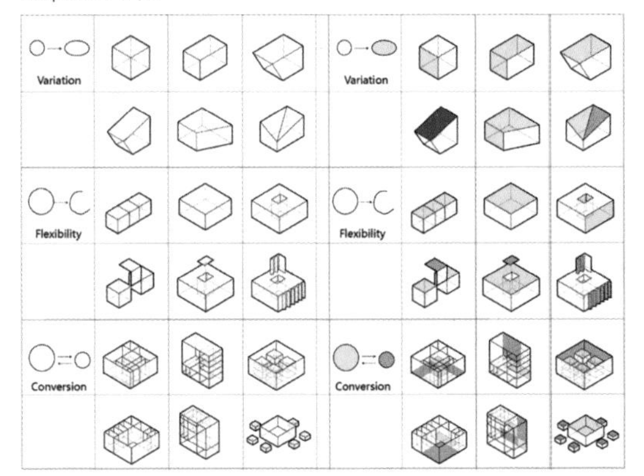


Hierarchy+Aggregation+Adaptation 层级 + 聚集 + 适应

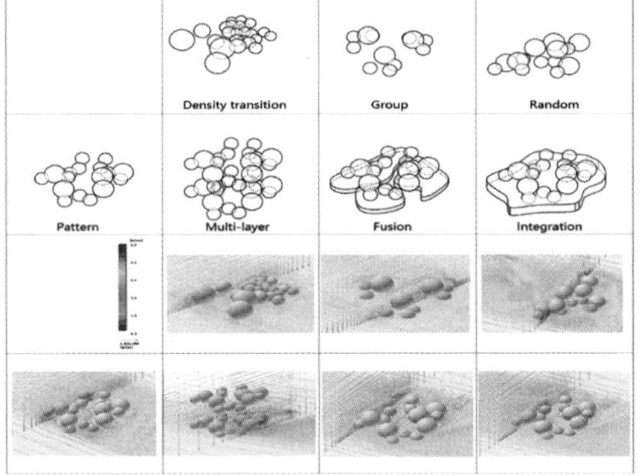


概念初期：关键词
Initial Concept: Keywords

以森林系统为主要研究对象，其热力学特性可抽象为三个关键词——聚集、层级、适应，分别对应自组织中协同作用产生的能量相互影响，能量在不同等级界面产生的耗散以及不断通过调节形态、交换能量等达到熵减少、增加有序度三个方面。

Taking the forest system as the main research object, its thermodynamic properties are abstracted into three key words—aggregation, hierarchy and adaptation. Corresponding to the mutual influence of the energy generated by the synergy in the self-organization, the energy is dissipated at different levels of the interface, and the entropy is reduced and the order degree is increased by adjusting the shape and exchanging the energy.

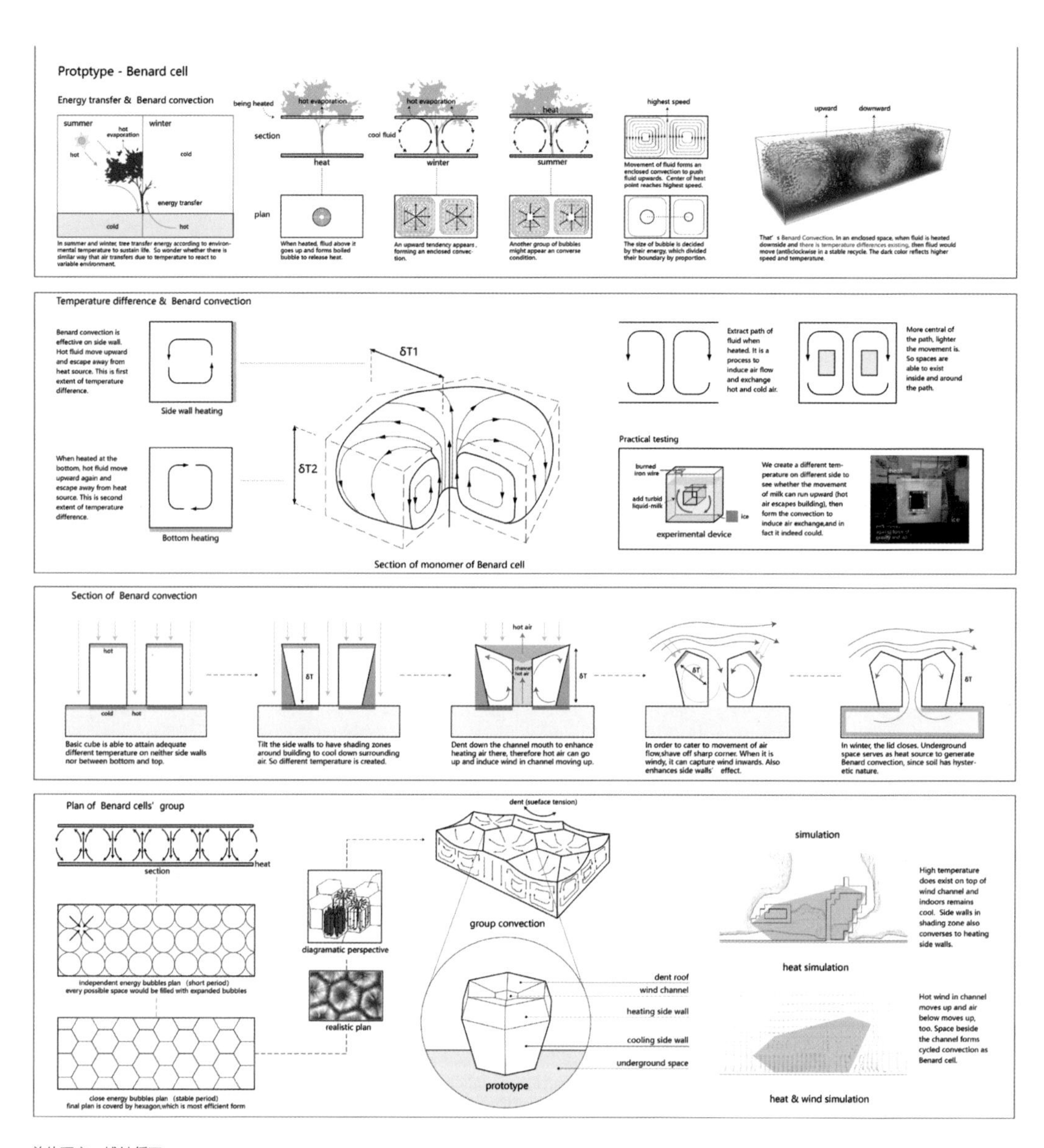

单体研究：博纳循环
Monomer Research: Benard Convection

单体树的能量流动符合物理学中的博纳循环，将循环流程抽象简化为两块有能量差的平板自发形成能量流动，流动路径之外的空间为建筑，而能量差则通过夏天集热与遮荫，冬天地暖与地上单体能量差实现。

The energy flow conforms to the Benard convection in physics, whose abstraction of the cycle process is simplified into two plates with energy difference spontaneously forming energy flow, and the space outside the flow path is the building. The energy difference is achieved by heat aggregating and shading in the summer, and temperature difference of underground and overground in the winter.

SINGLE 单体

AGGREGATION 群体

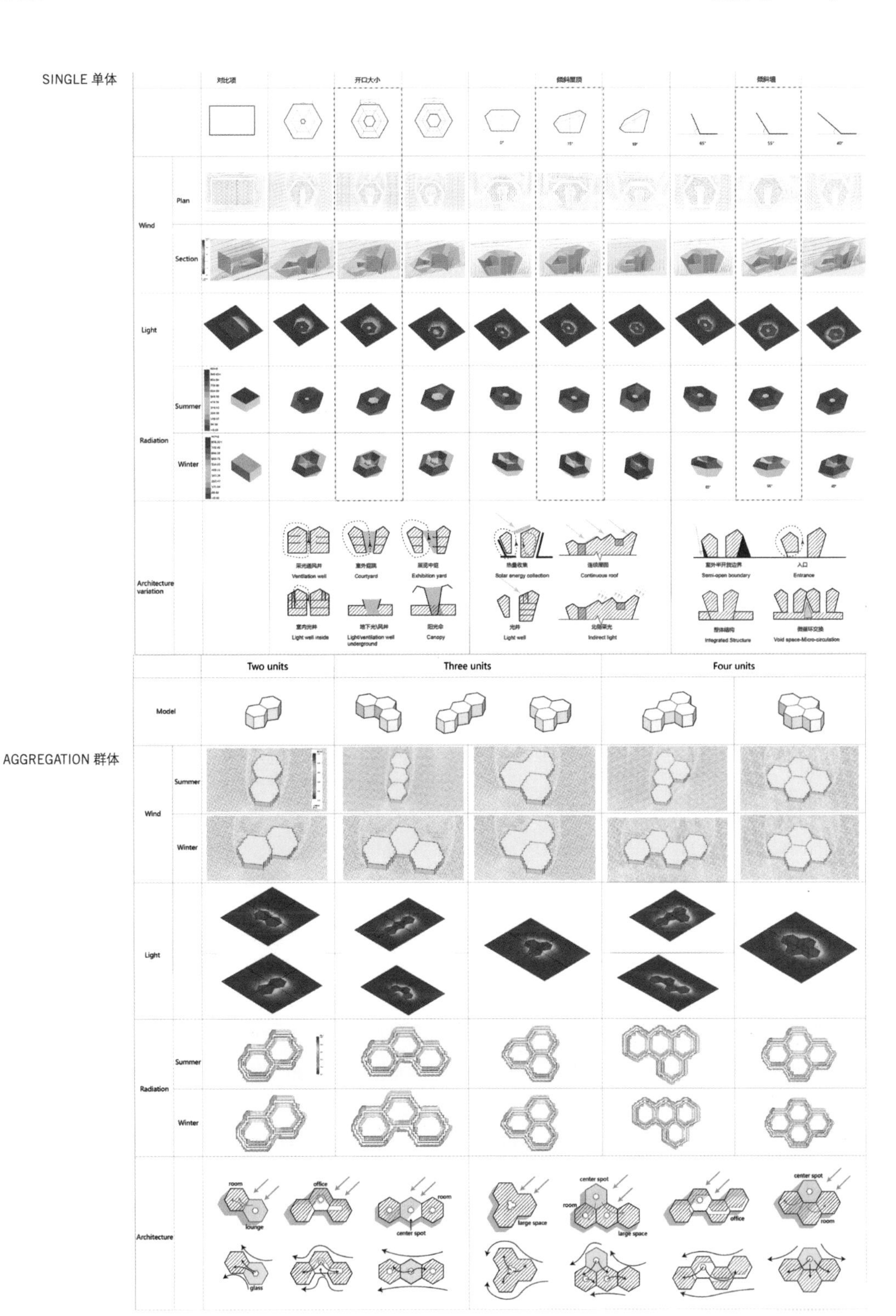


模拟优化

Simulation & Optimization

## 热力学建筑原型 Thermodynamic Architectural Prototype

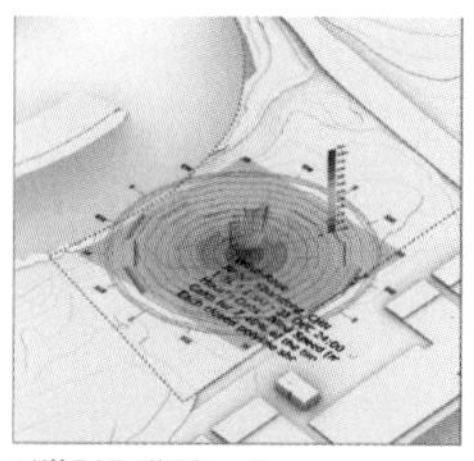

1.基地最主导环境因素——风
The most dominant element in site-wind forms the shape firstly.

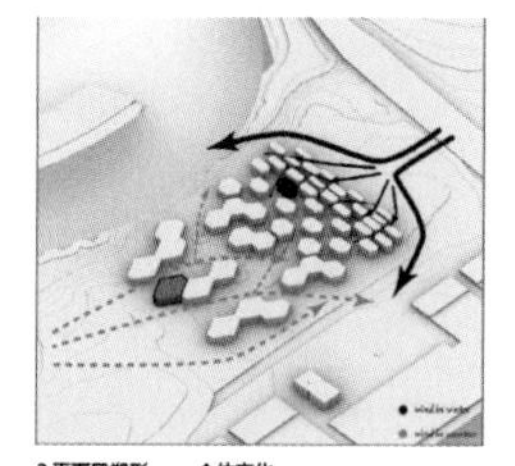

2.平面风塑形 ——个体变化
——群体疏密变化
Wind morphing——Single shape variation
——Group density variation

3.平面塑形优化——Pattern优化
Masterplan arrangement optimization——Pattern Optimization

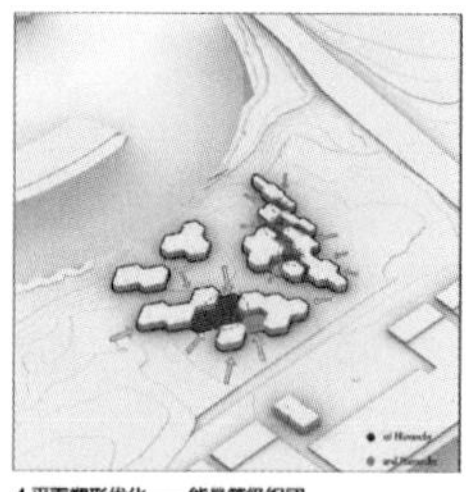

4.平面塑形优化——能量等级组团
Masterplan arrangement optimization——Energy Hierarchy Optimization

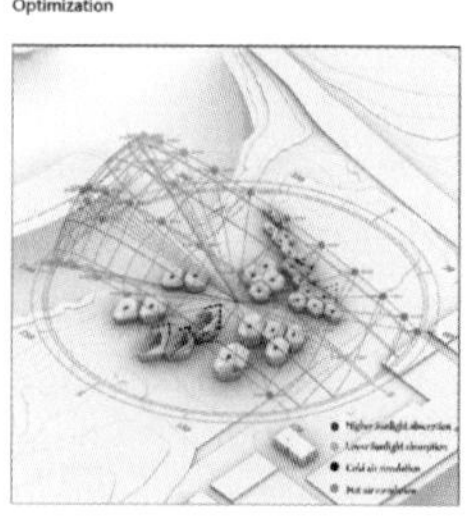

5.次主导光和热的介入——能量吸收最大化产生伯纳循环
Second dominant element——light & radiation
Maximization of energy absorption to generate benard circulation

6.风光热协同优化
Synergistic optimization of wind and solar energy

7.季节性能量转移
Seasonal energy transfer

8.室外微环境——基地原有资源(树、水)的植入再优化
Outdoor microclimate——implantation of resourses in the site(Trees & River)

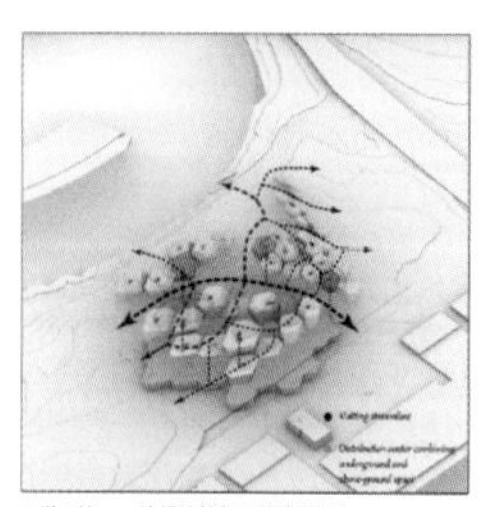

9.微环境——流线功能与环境的再整合
Microclimate——Integration of environment, circulation and function

10.室外微环境——室内外缝隙空间的植入
Outdoor microclimate——Outdoor void space implantation

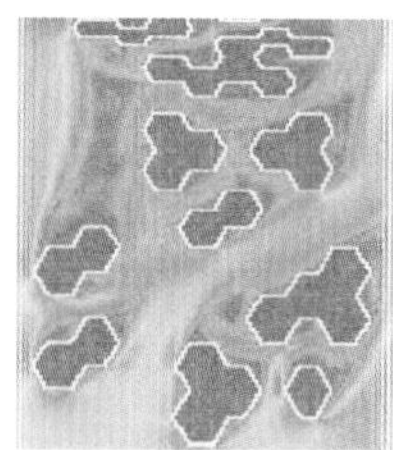

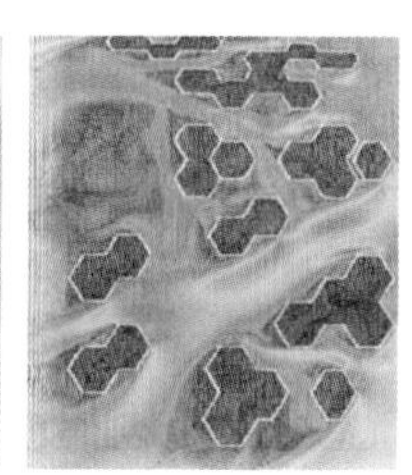

第一稿：顺风，缺少回流与阻力，风道单一

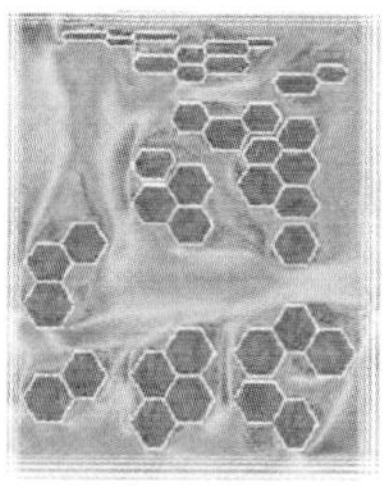

第二稿：疏导风，加入单体间组合的大伯纳胞，不同尺度呼应原型

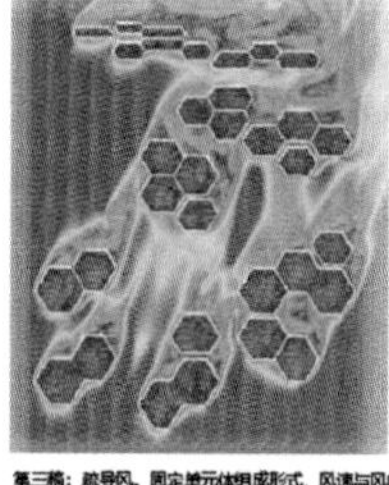

第三稿：疏导风，固定单元体组成形式，风速与风向控制欠缺

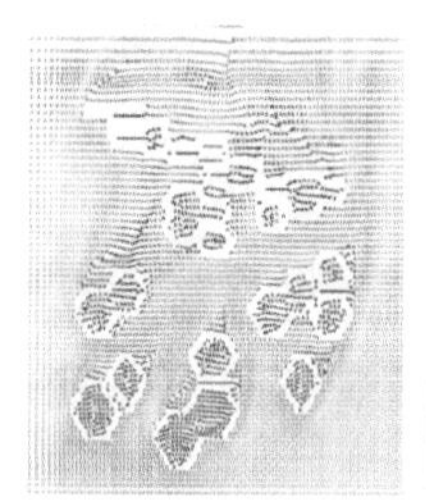

第四稿：疏导风，优化北边缘状单元体，增加与主体联系

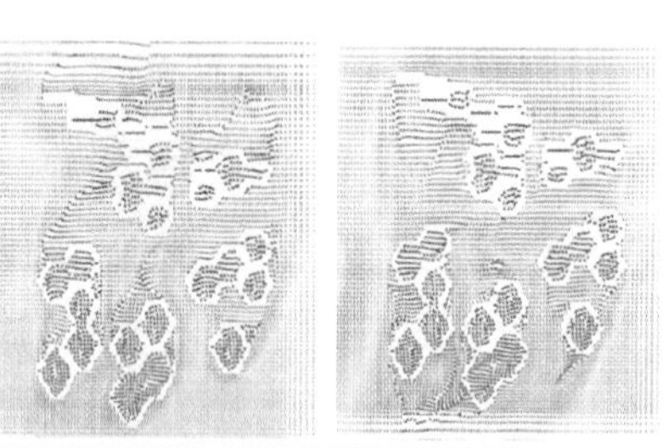

第五稿：疏导风，适当减少河水、大片草处风速，固定单元形式，贴合人体舒适度

设计生成
Design Generation

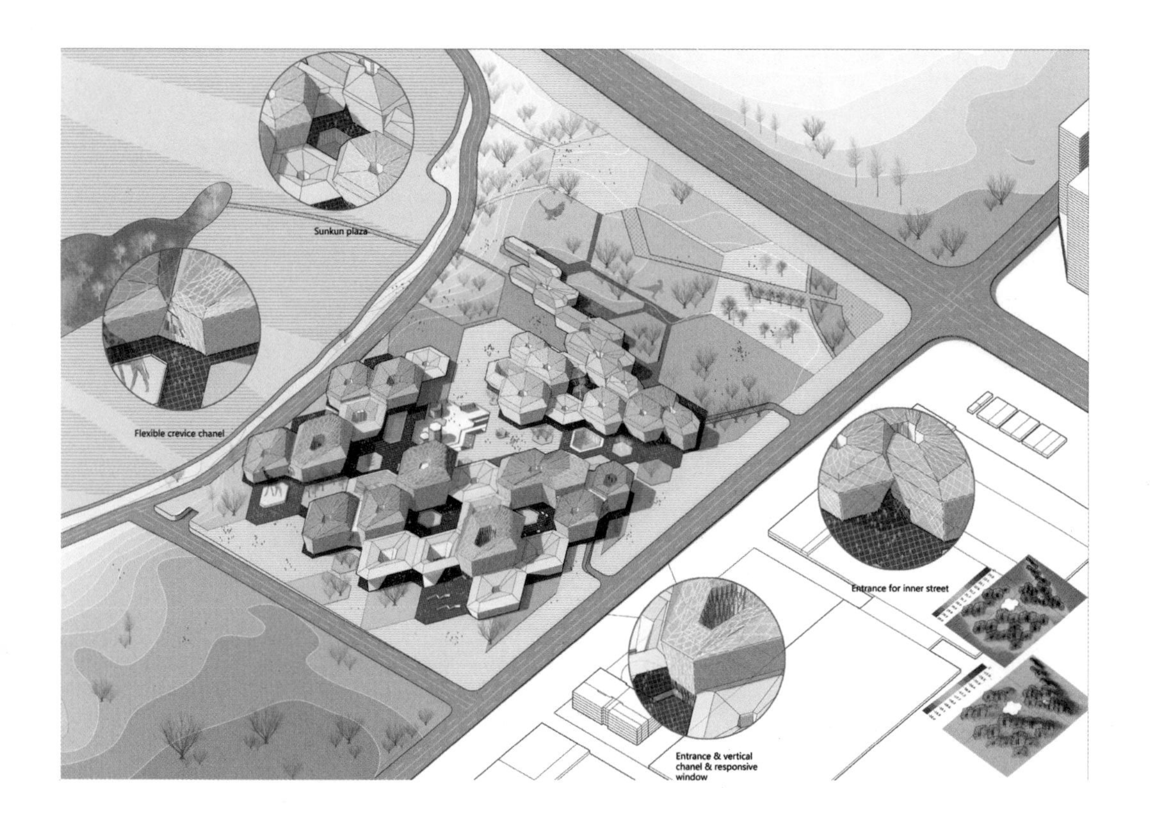


生成：类比森林演替在不同阶段引入不同元素，方案生成按要素重要性先后引入风、光、热等自然资源和流线功能等，最终形成不同能量等级的热力学森林公园。

Generation: Analogous forest succession introduces different elements at different stages. The program generates natural resources and streamline functions such as wind, light and heat according to the importance of the elements, and finally forms thermodynamic forests with different energy levels.

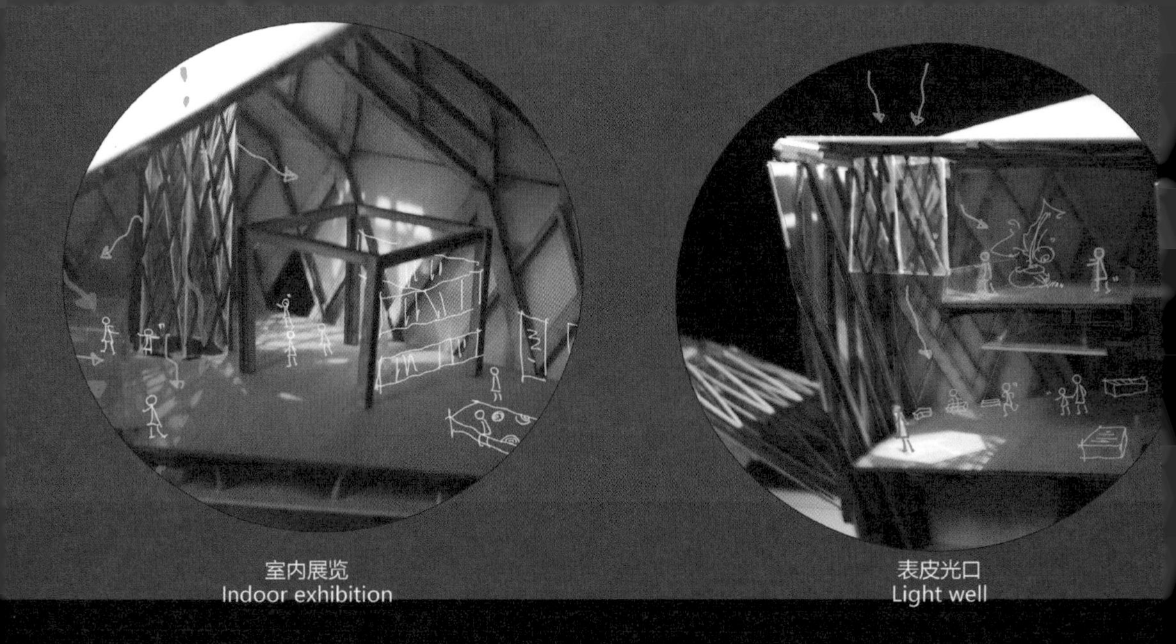

室内展览
Indoor exhibition

表皮光口
Light well

建筑选择木结构作为主体结构。一方面考虑当地编织文化，另一方面木材保温性能良好，多用于同纬度地区的建筑。

We chose the wood structure as the main structure. On the one hand, we considered the local weaving culture and wanted to make it into use in our design, on the other hand, the wood insulation performance is good, and it is mostly used in buildings in the same latitude area.

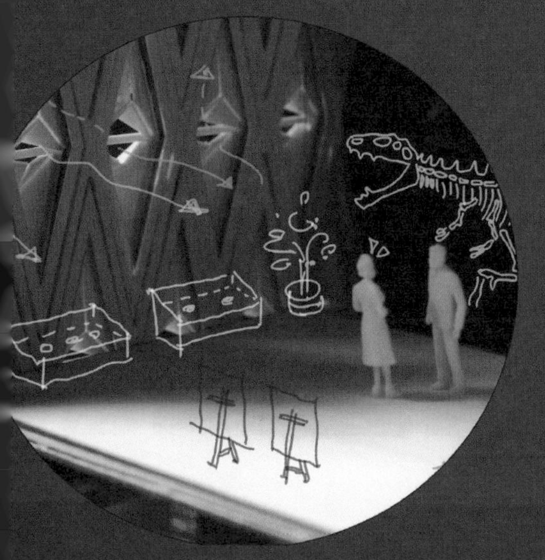
可呼吸表皮
The flexible epidermis

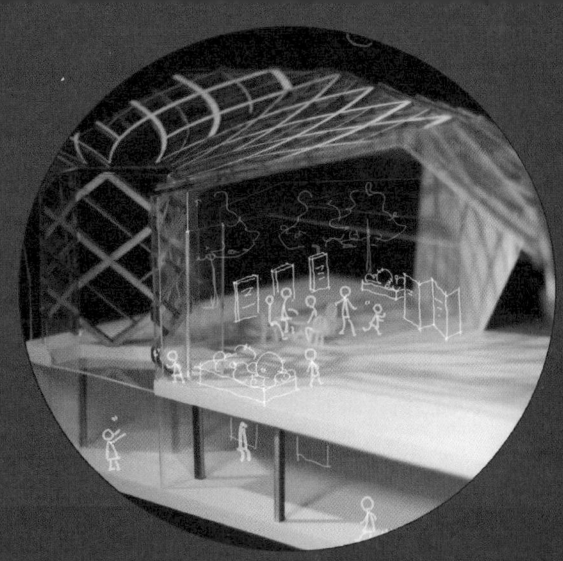
室外展厅
The flexible epidermis

单体剖模型

### 通过叶片开合控制气孔大小
Dynamic Facade

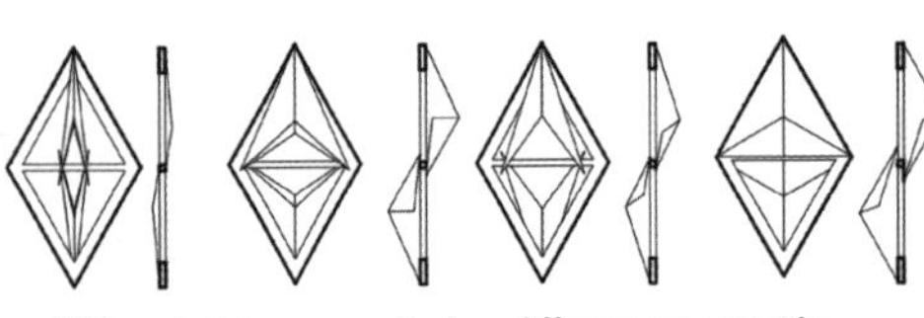

Different states are applied on different seasons. The hotter the weather is, the bigger the holes are.

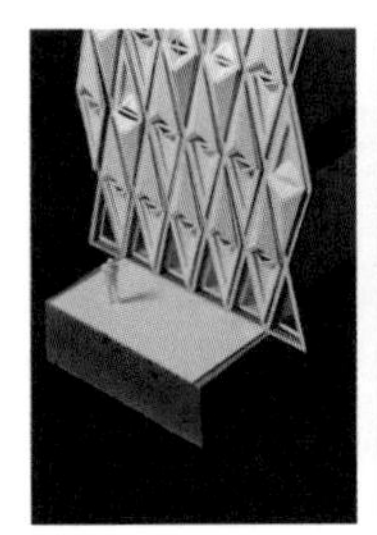

建筑内经过组织的气流在单体内部筒体中沿垂直方向流动，且气流出入口位于单一立面，所以立面构件需要保证气流有进出建筑两个方向，彼此相反，且气流方向根据室内功能有一定调整，但大体入口在下部，出口在上部。

为了满足季节性需求，开口设计为可开合构件：夏季越炎热，构件开口角度越大；冬季越寒冷，角度越小，极端情况完全关闭。

The organized airflow in the building flows vertically in the inner cylinder of the unit and is located on a single facade. Therefore, the wind must enter and exit the building in two directions, opposite to each other, and the direction is adjusted according to the indoor function, but the general entrance is in the lower part, the outlet is in the upper part.

In order to meet seasonal needs, the opening is designed to be openable and closable. When the summer is hotter, the opening angle of the component is larger. When the winter is colder, the angle is flatter and the extreme situation is completely closed.

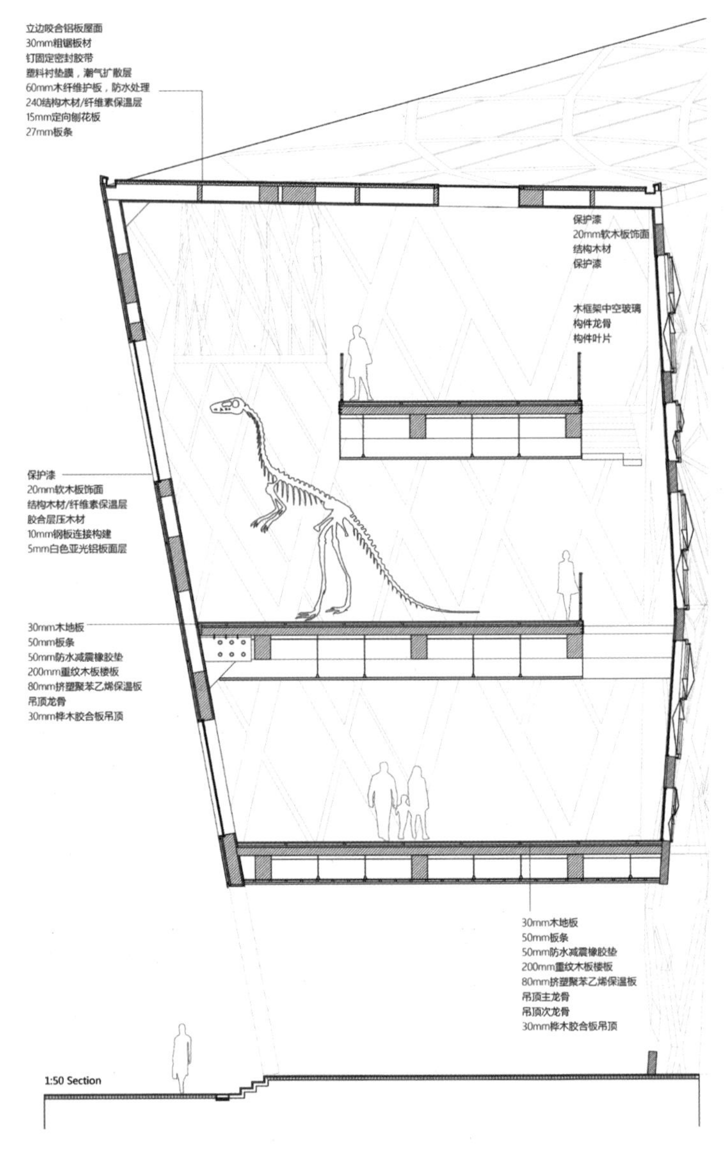


细部设计
Detail Design

夏
Summer
冬
Winter

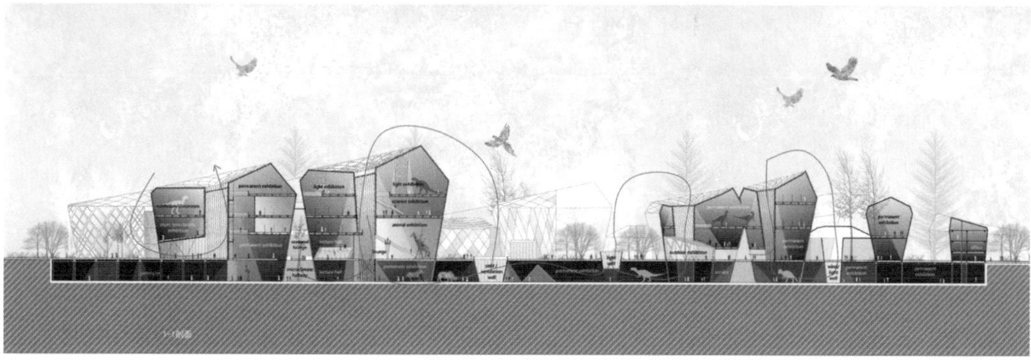

我们小组提出的原型概念为“双系统空气流动——作为交换机器的可适应性建筑”。我们选取应对极端气候的四个生物器官原型：海豹的鼻腔、金枪鱼的鳍、羚羊的脑部、企鹅的脚。通过研究其共性与机制，我们发现它们都是通过交换来实现对极端气温的应对的。通过引入第二套系统，两套系统之间进行流动，进而实现交换。最终我们根据生物原型提取出三个关键词：梯度、流动、交换。并以此形成建筑原型的机制：通过创造梯度来形成流动，通过流动来实现交换，以应对极端环境。

The prototype concept proposed by our team is “Bi-sourse Reactivity—Adaptive Architecture as an Exchange Machine.” We selected four biological prototypes for extreme climate: the elephant seal's nasal cavity, the tuna's fin, the antelope's brain and penguin's foot. By studying their commonalities and mechanisms, we find that they all use exchange mechanism to achieve the adaption to extreme temperature. By introducing a second system, the two systems flow between them achieve exchange. In the end we extracted three keywords based on the biological prototype: gradient, flow, exchange. And to form the mechanism of the building prototype: the flow is created by creating a gradient, and the exchange is realized by the flow to cope with the extreme environment.

# 双系统空气流动
# Bi-sourse Reactivity

学生 Students：
黄景溢 HUANG Jingyi
莫然 MO Ran
陈昌杰 CHEN Changjie
克里斯托夫・芬克 Kristoff Fink

指导老师 Tutor：
李麟学 LI Linxue

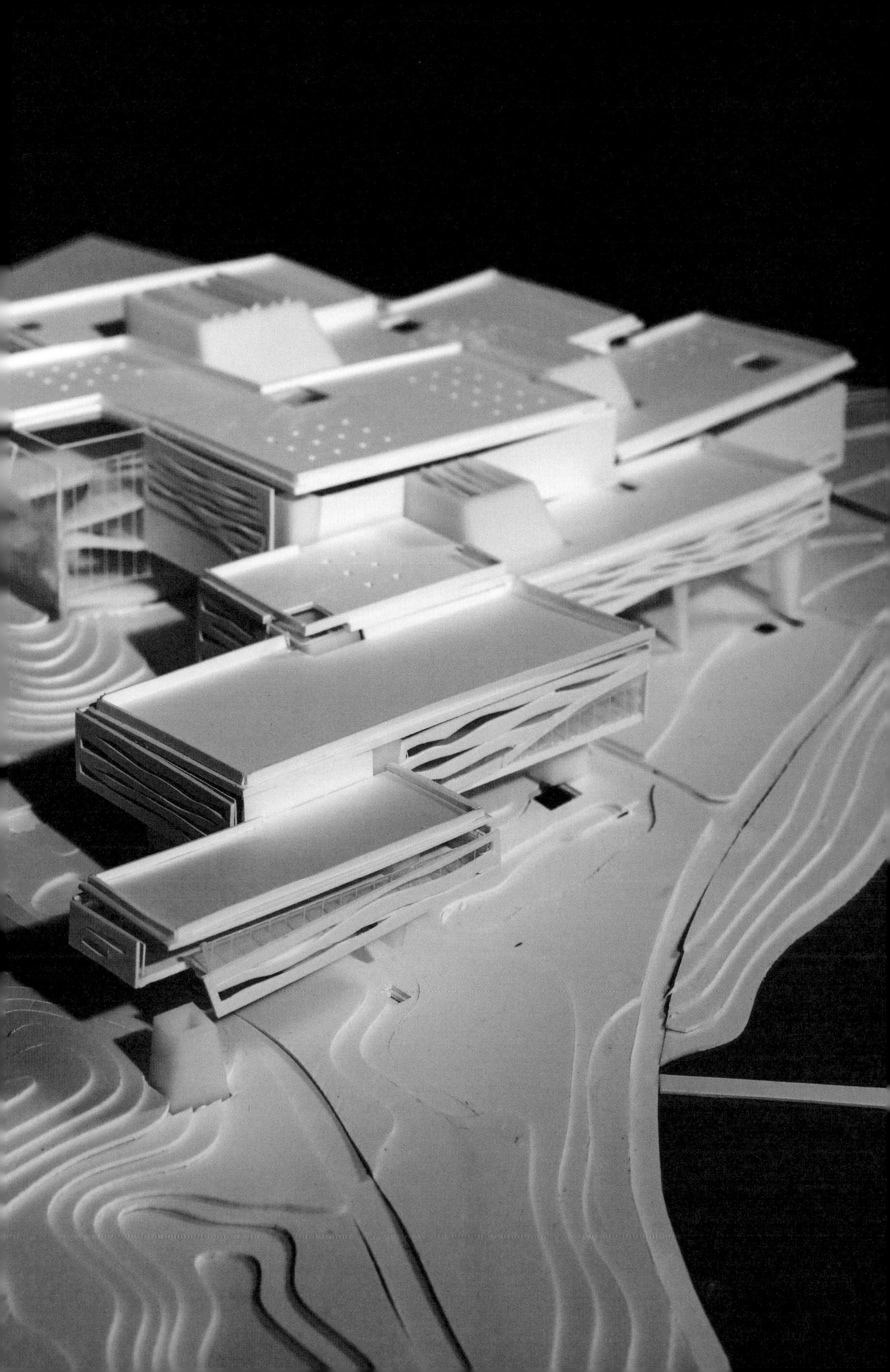

## 热力学建筑原型 Thermodynamic Architectural Prototype

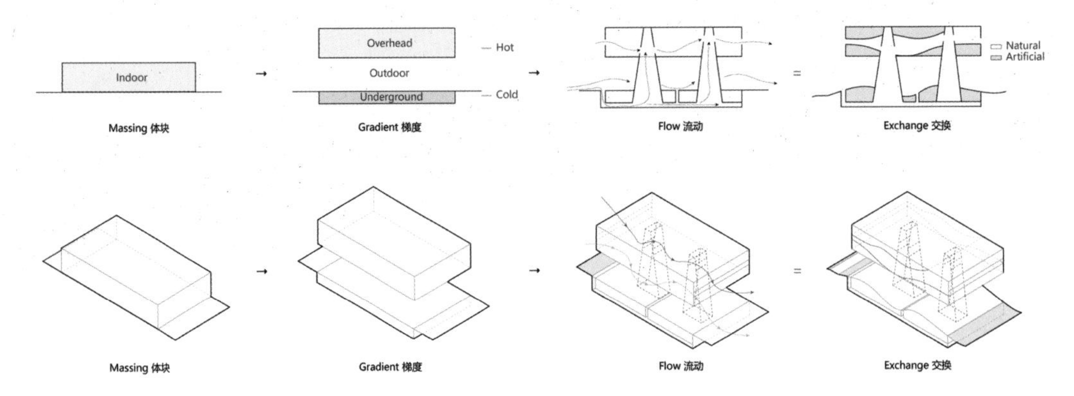


生成 Generation

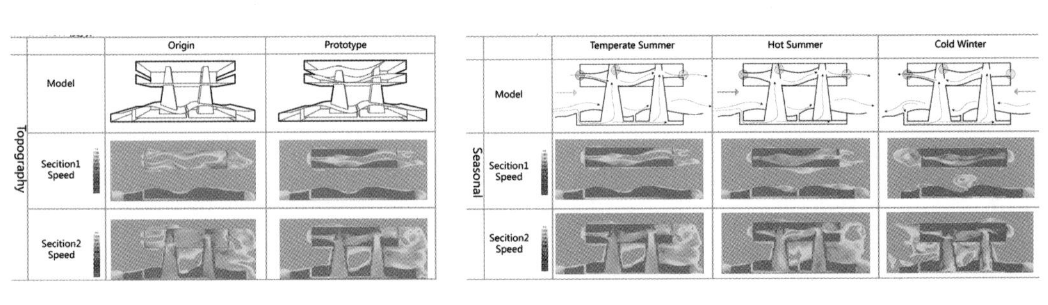


模拟 Simulation

季节性适应 Seasonal Adaption

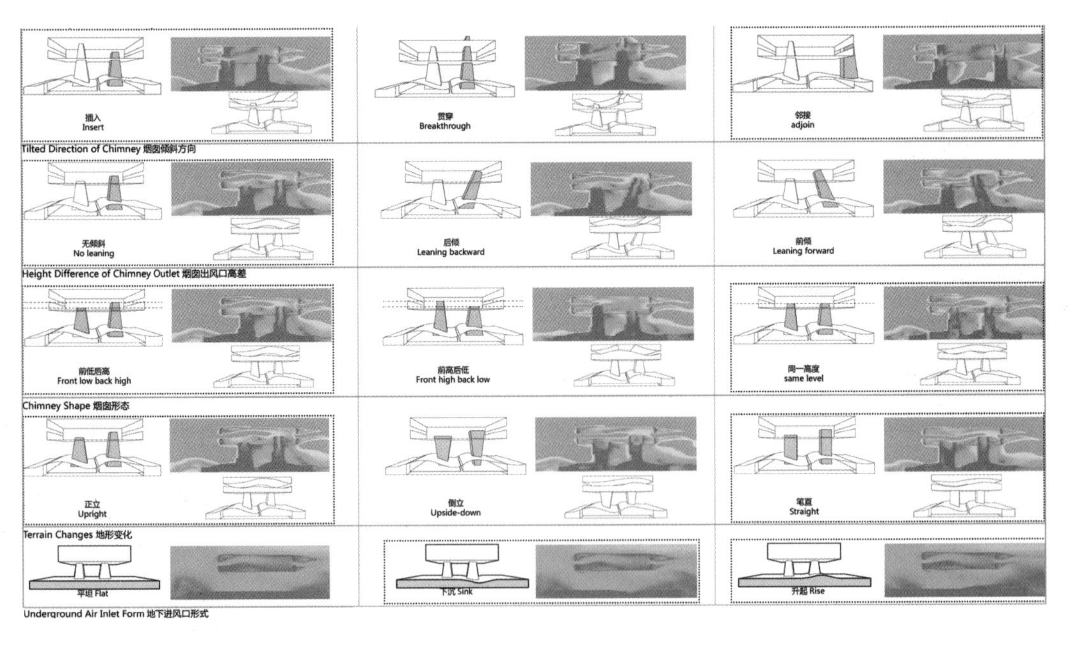


原型元素研究 Prototype Element Research

原型生成

Prototype Generation

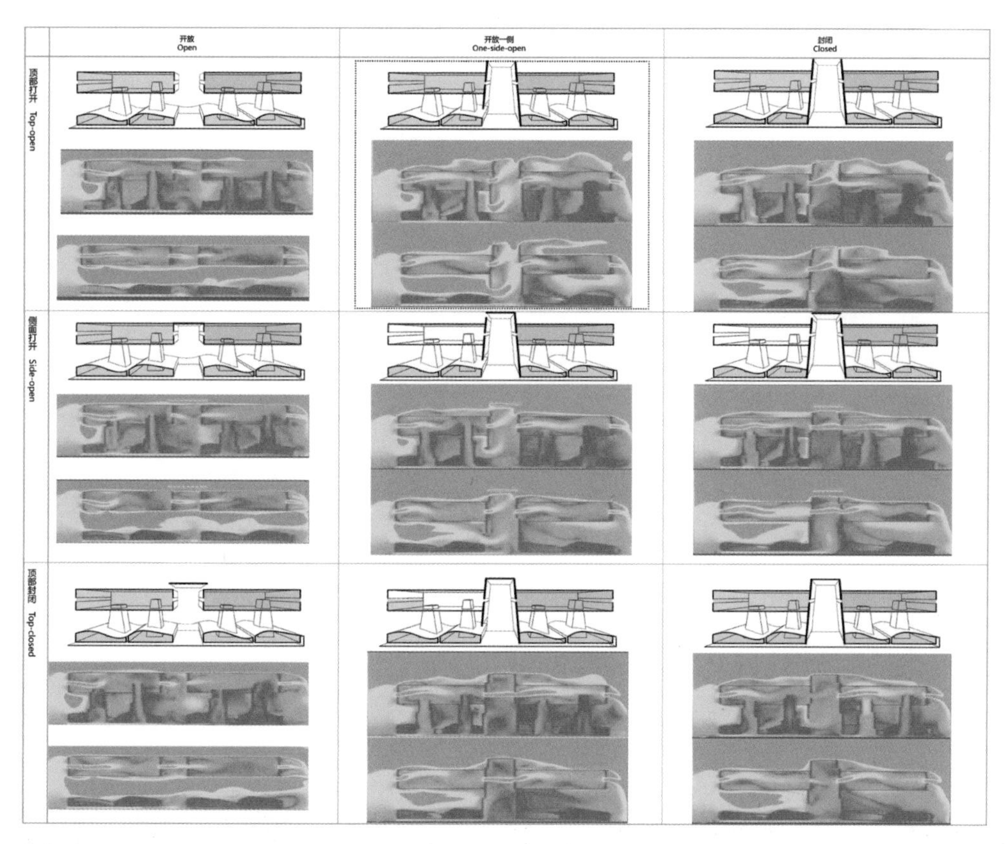


庭院形式 Courtyard

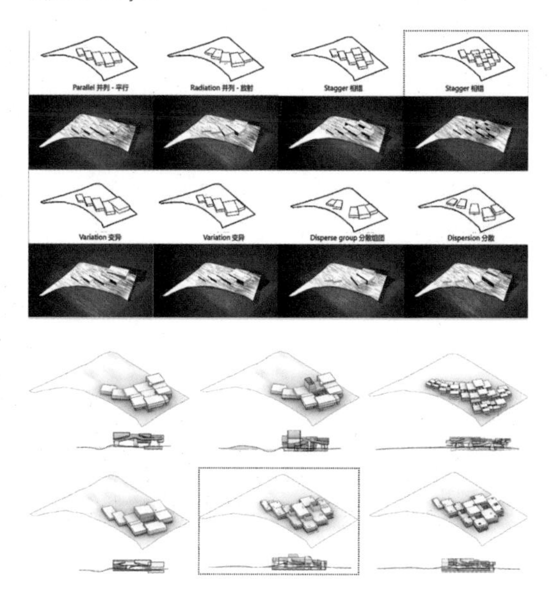


生成研究 Generation Research

原型研究
Prototype Generation

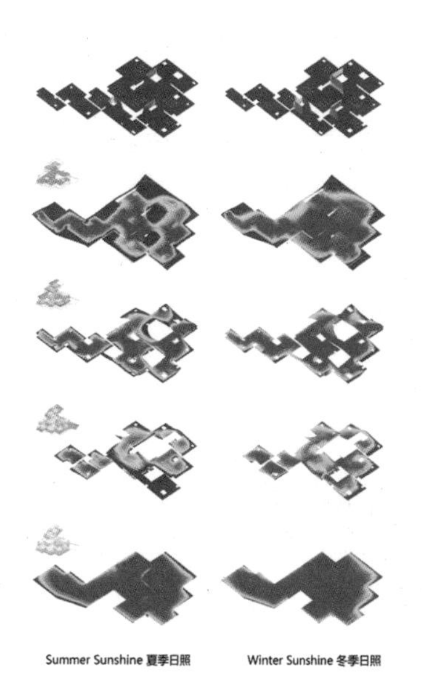


整体辐射验证 Overall Verification-Radiation

## 热力学建筑原型 Thermodynamic Architectural Prototype

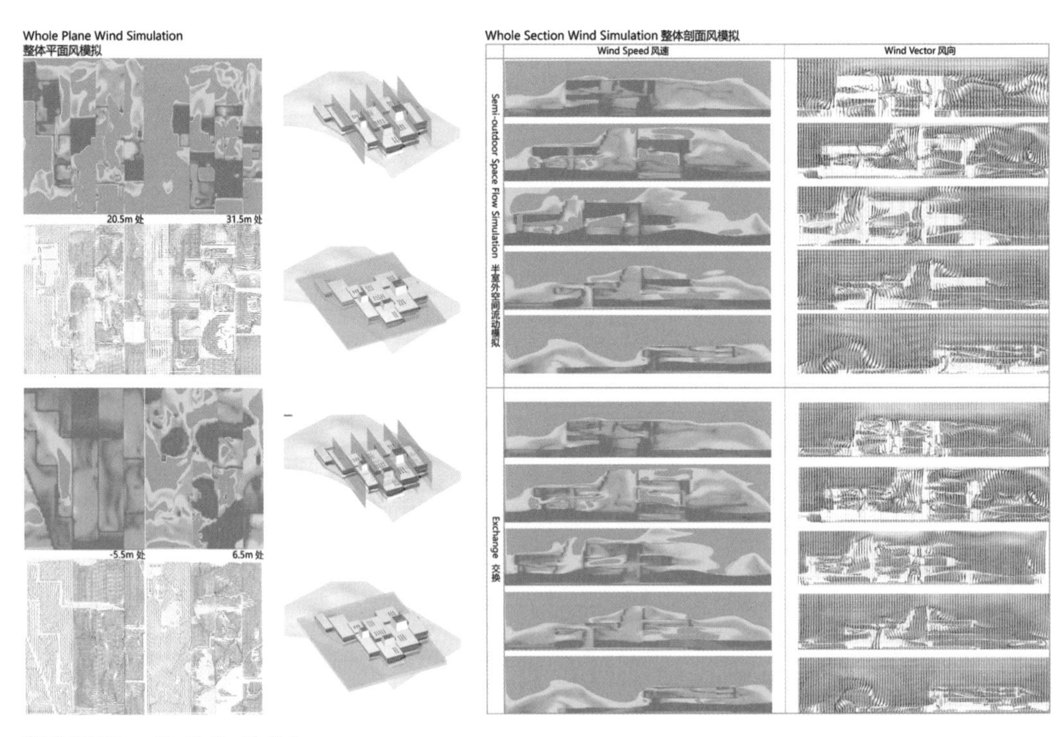


整体通风验证 Overall Verification-Ventilation

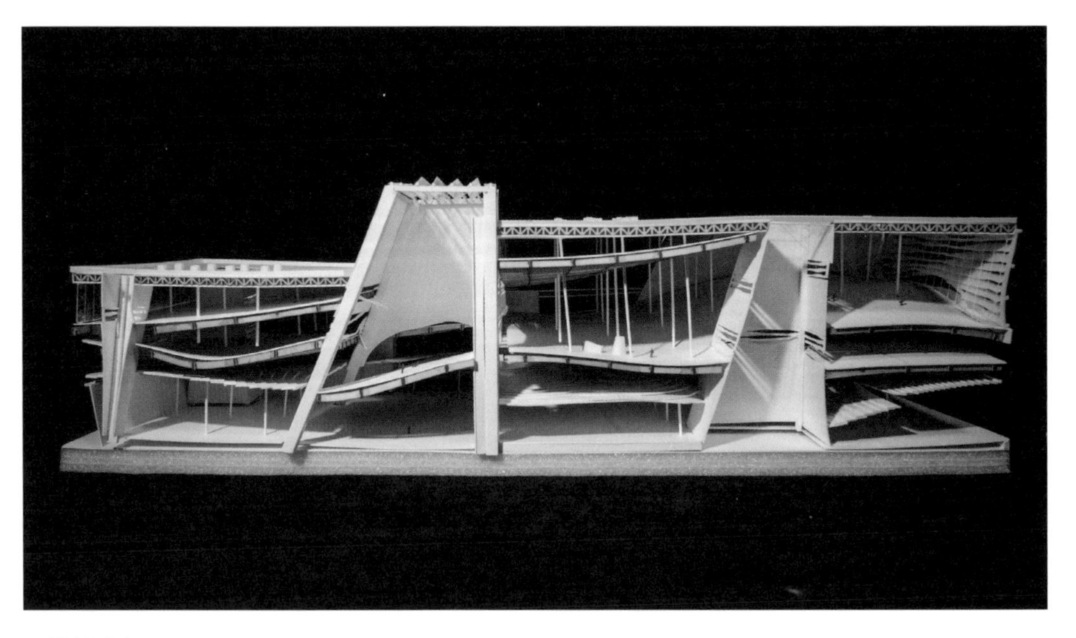

双系统空气流动
Bi-source Reactivity

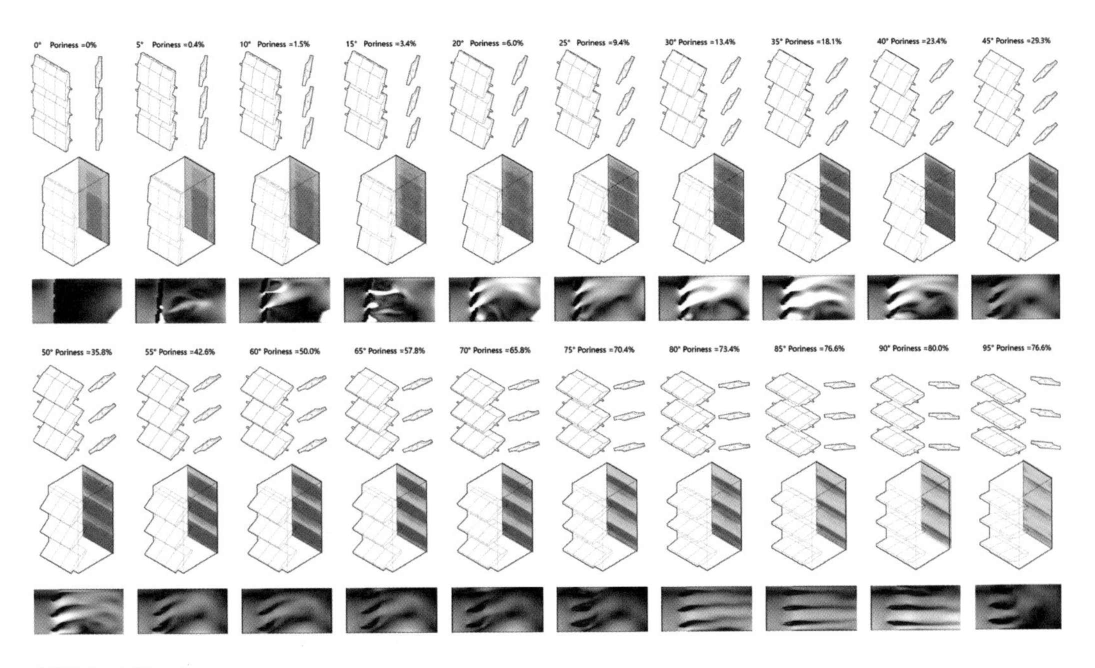


孔隙率 Material Porosity

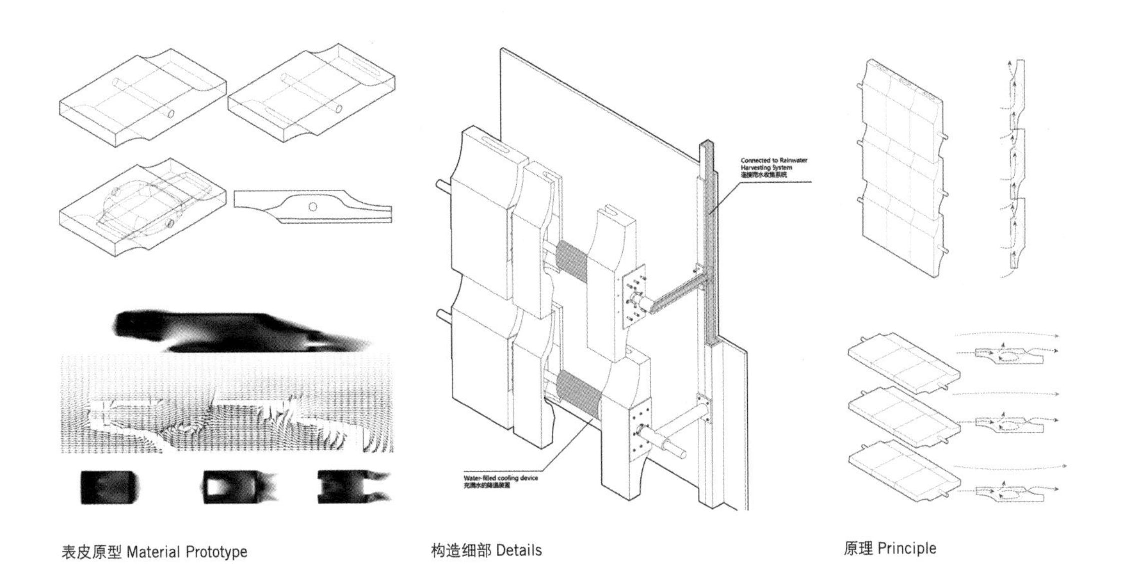


表皮原型 Material Prototype

构造细部 Details

原理 Principle

材料文化
Material Culture

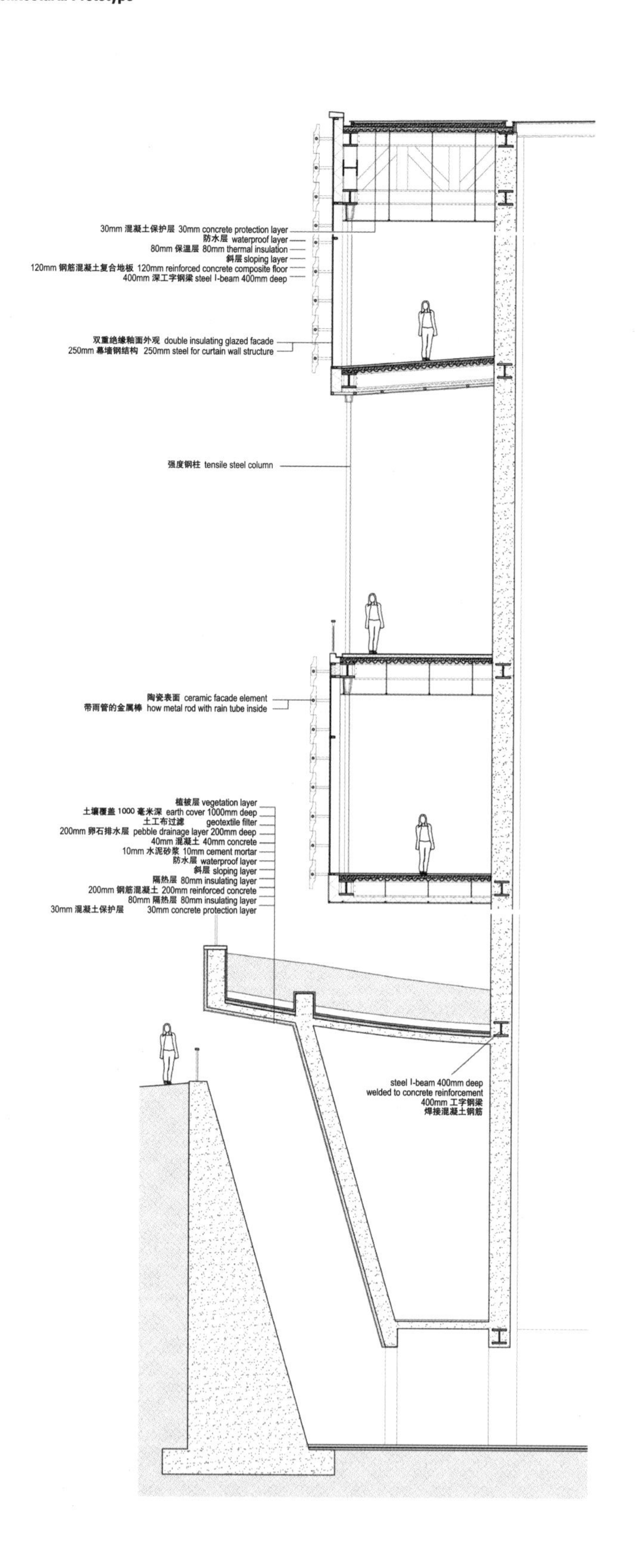


构造详图

Construction Detail

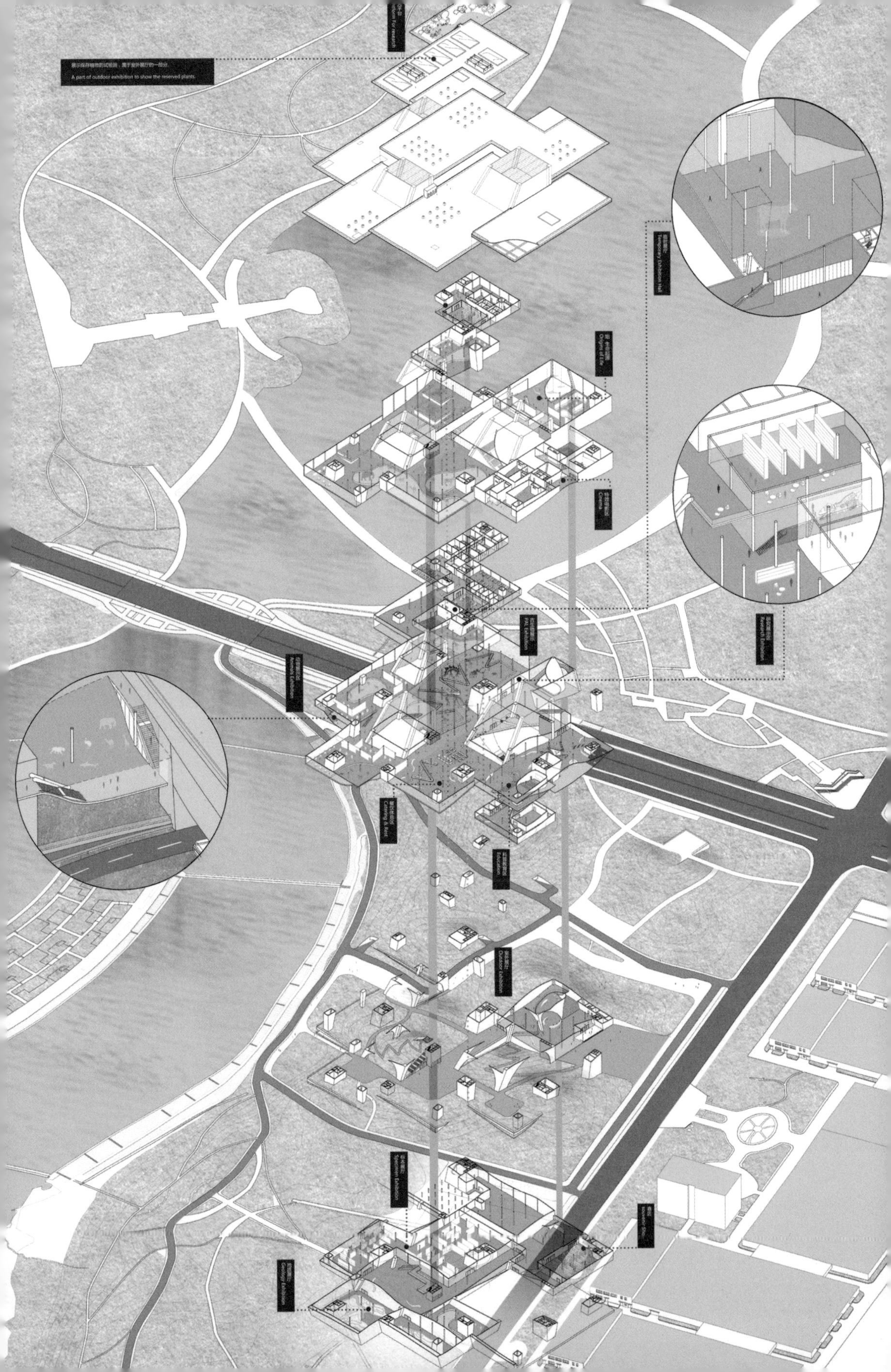

A part of outdoor exhibition to show the reserved plants.
Temporary Exhibition Hall
Origins of Life
Cinema
PAL Exhibition
Research Exhibition
Animals Exhibition
Catering & Rest
Education
Outdoor Exhibition
Specimen Exhibition
Geology Exhibition

在生物界中，有一项造物是符合我们追求的内稳态的——那就是白蚁穴。白蚁穴通常位于热带地区，但却可以在日光暴晒下保持内部凉爽通风，同时夜间不至于过冷。在研究了它的运作规律之后，我们提取了关键词——空腔、孔隙与温室。空腔形成洞穴为建筑提供了缓冲，孔隙使内外气体流通，温室嵌入体块成为建筑的热源。建筑的形态是一个多孔的岩石。通过孔隙和腔体的控制作用，建筑能实现冬季能量的最大输入和夏季风向的最大渗透。此外，温室作为调节器以维持建筑的稳定状态。

In the field of biology, there is a creature that is consistent with our target: the termite mound. They are usually located in tropical areas, but they keep the inside cool and ventilated well. After studying its operation, we extracted the key words – cavity, porosity, and greenhouse. The cave formed by the cavity provides a buffer for the building, the pores allow the inside and outside of the gas to circulate, and the greenhouse embedded in the block becomes the heat source of the building. The shape of the building is a long and short rock that can achieve maximum input of winter energy and maximum penetration of summer wind direction through the control of pores and cavities, and the greenhouse acts as a regulator to maintain steady state.

# 空腔 · 孔隙 · 温室

# Cavity, Porosity, Greenhouse

**学生 Students：**
**薛钰瑾 XUE Yujin**
**刘帆 LIU Fan**
**胡雨 HU Yu**
**布拉德利 · 艾伦 · 埃勒布雷克特 Bradley Allen Ellebracht**

**指导老师 Tutor：**
**李麟学 LI Linxue**

夏季 Summer

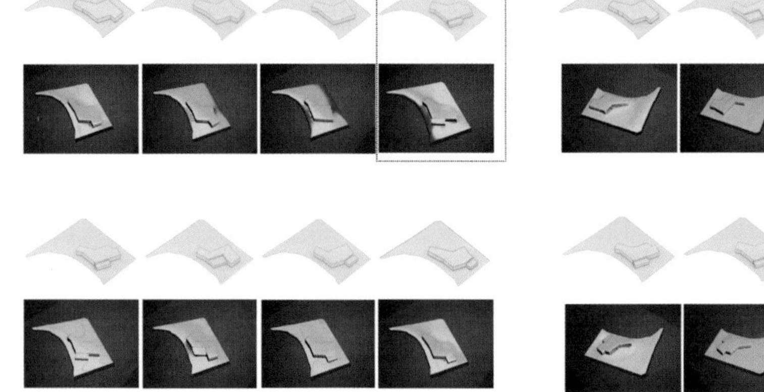

冬季 Winter

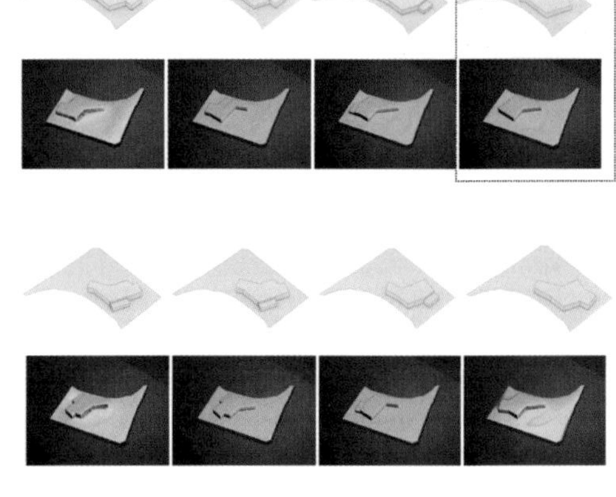

风压 Wind Pressure

环境分析
Environment Analysis

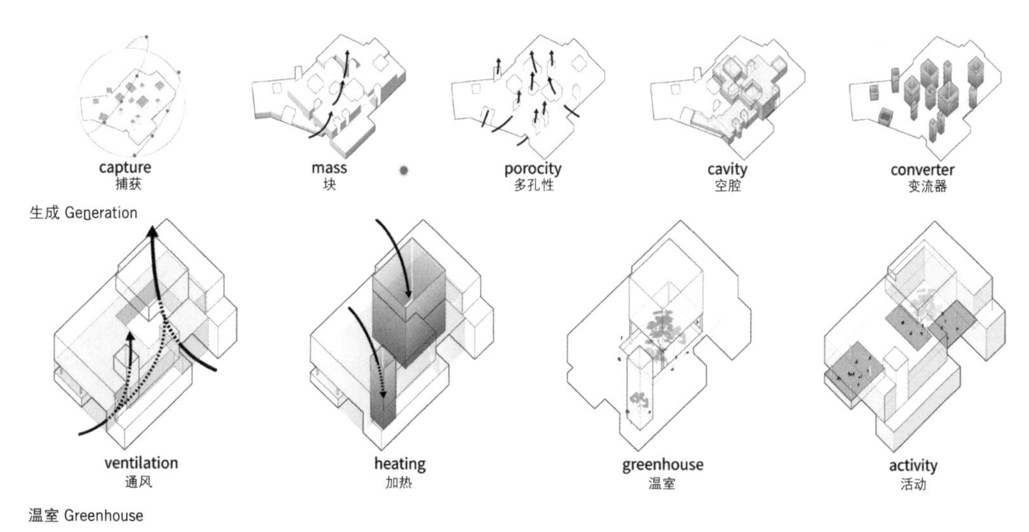


概念生成
Concept Generation

## 热力学建筑原型 Thermodynamic Architectural Prototype

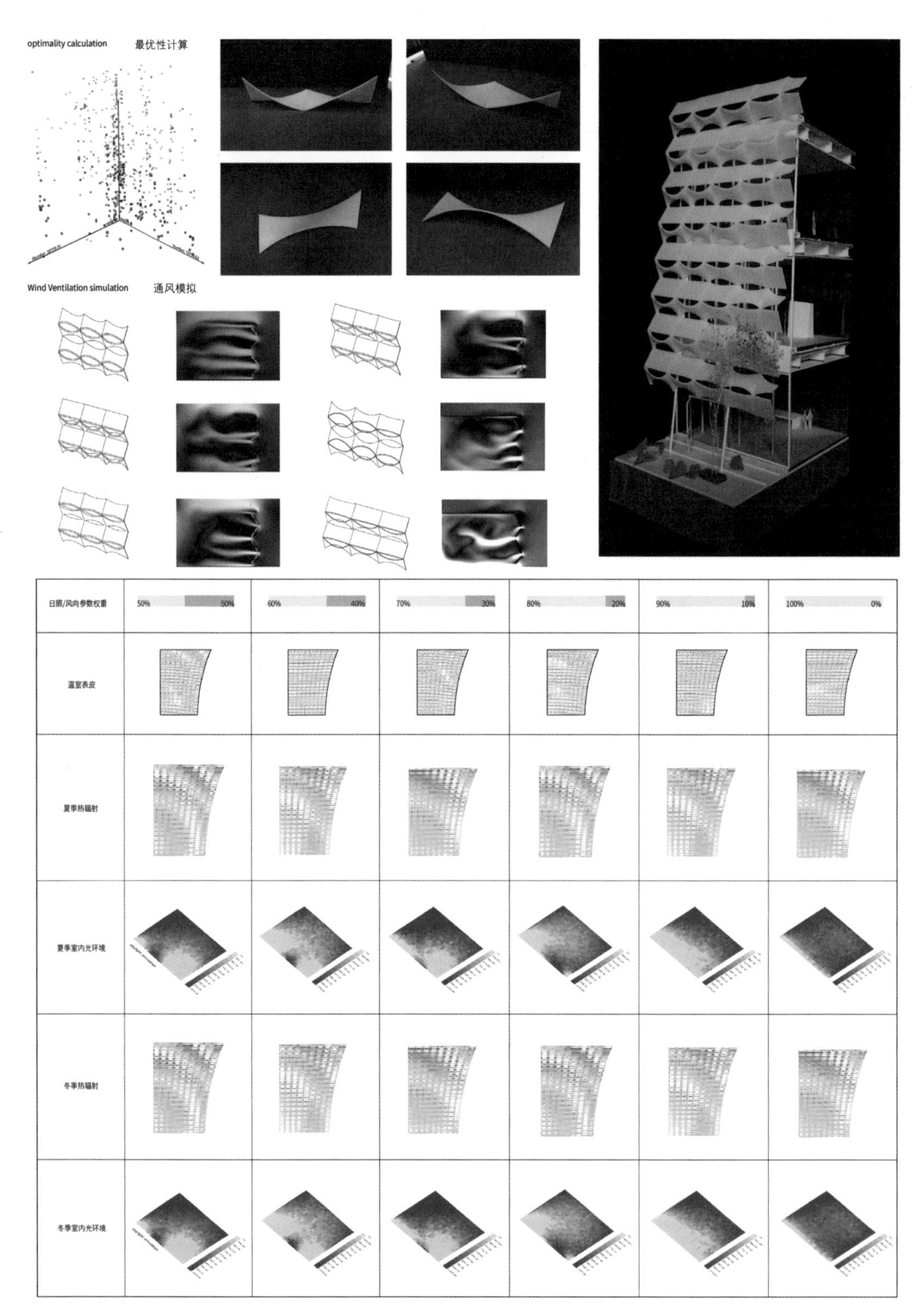

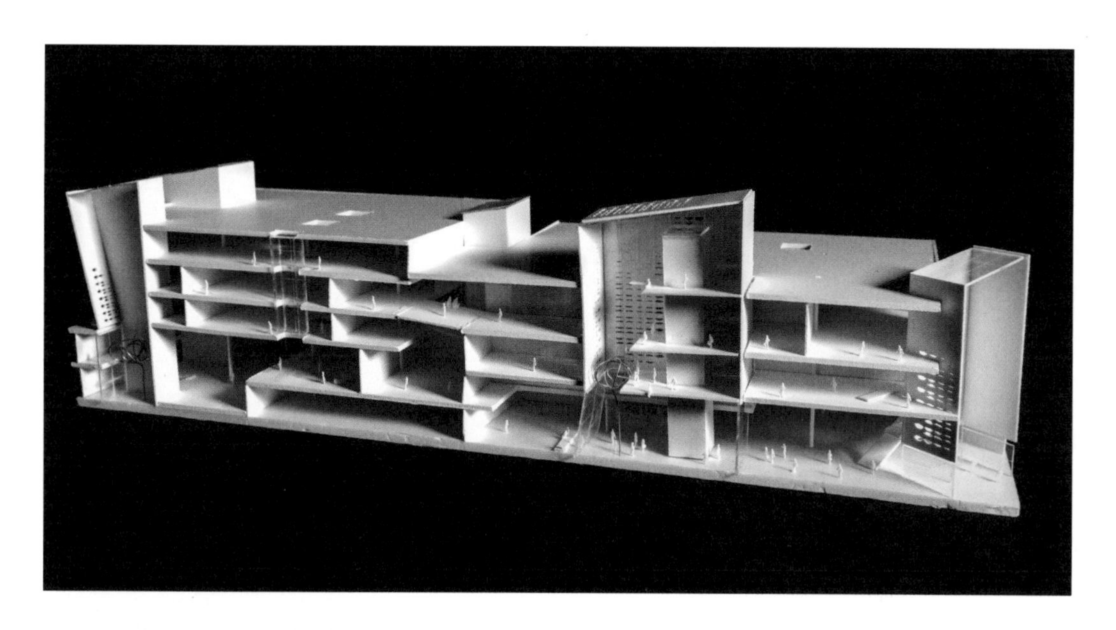

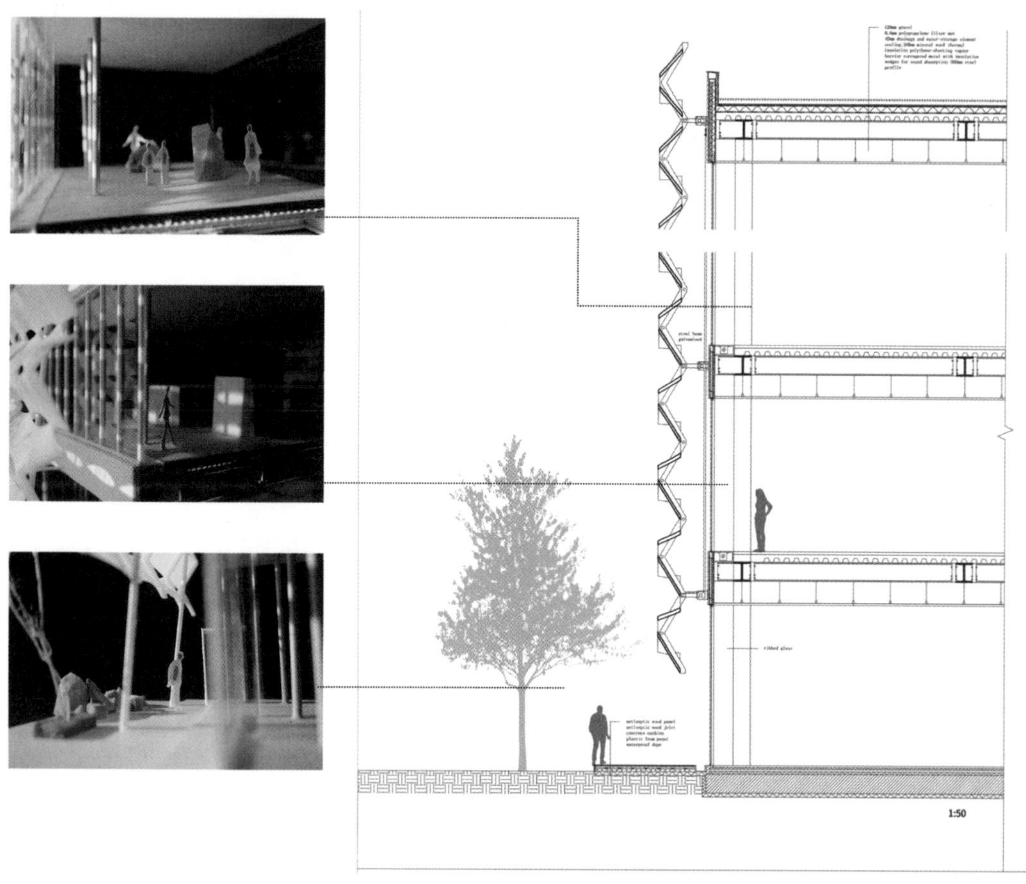


细部设计
Detail Design

Wind Roads
Arterial Roads

# 亚洲垂直城市
# Vertical Asia Cities

新加坡作为一个城市国家，土地资源有限，面临着未来持续增加的人口压力。基地巴耶利巴历史上曾经作为民用机场，樟宜机场建成后，转为军用机场。场地总面积 6.78 平方公里，设计要求在其中选择 1 平方公里的地块作为基地设计 10 万人居住的城市。基地内含有一条机场跑道，基于飞机起飞时的空气动力学原理，机场跑道方向与新加坡主导风向一致。场地中央有一大片森林沿着机场跑道展开，成为非常重要的自然资源。

As a city country, Singapore has limited land resources and is facing the pressure of a growing population in the future. The base of Paya Lebar served as a civilian airport. Before being converted into a military airport when Changi Airport was built. The total area of the site is 6.78 square kilometers. The design requires a plot of 1 square kilometer to be used as a base to design a city for 100,000 people. The base contains an airport runway whose direction is same as local dominant wind to fit the aerodynamics when aircrafts take off. There is a large forest in the center of the site that runs along the airport runway and becomes a very important natural resource.

在新加坡的巴耶利巴，我们关注能量流，并设想了一个基于当地气候和文化的热力学城市范式——“雨林城市”。通过自然元素、自组织系统、群落结构和城市演替的整合，每个人都会栖息在热带雨林的植物和动物群中。就这样，我们将创建一个人人共享的城市。

“雨林城市”是一种新的城市模式，正如法国哲学家拉图尔呼吁的那样，更紧密地联系自然与社会，以迎接未来城市的挑战。在雨林城市，我们以一种小的、分散的、聚集的方式参与自然、工业、社区等的能量流动。这是一个给人完全不同于以往体验的城市，一个人人共享和贡献的城市。

In Paya Lebar, Singapore, we focused on energy flow and imagined a thermodynamic urban paradigm based on the local climate and culture, a “Rainforest City”. Through the integration of natural elements, self-organization system, coenosium structure and city succession, everyone will dwell poetically there with the flora and fauna in the tropical rain forest. Just like this, we will create a city where everyone contributes to flow and vitality of the shared system .

Rainforest City is a new city model, as French philosopher Latour appeals, a closer connection between nature and society to meet the future urban challenge.In the Rainforest City, we participate the energy flow of nature, industry, community, etc. in a small, scattered and gathering way. We would experience an entirely different city where everyone shares and contributes.

# 雨林城市
# Rainforest City

学生 Students：
夏孔深 XIA Kongshen
范雅婷 FAN Yating

指导老师 Tutor：
李麟学 LI Linxue

指导老师：李麟学、姚栋、黄一如
评审老师：CJ Lim、Ada Fung Yin-Suen、Vishaan Chakrabarti、Chan Sau Yan
学生：范雅婷、方荣靖、韩雪松、邝远潇、刘晗、夏孔深
Tutors: LI Linxue, YAO Dong, HUANG Yiru
Review Teachers: CJ Lim, Ada Fung Yin-Suen, Vishaan Chakrabarti, Chan Sau Yan
Students: FAN Yating, FANG Rongjing, HAN Xuesong, KUANG Yuanxiao, LIU Han, XIA Kongshen

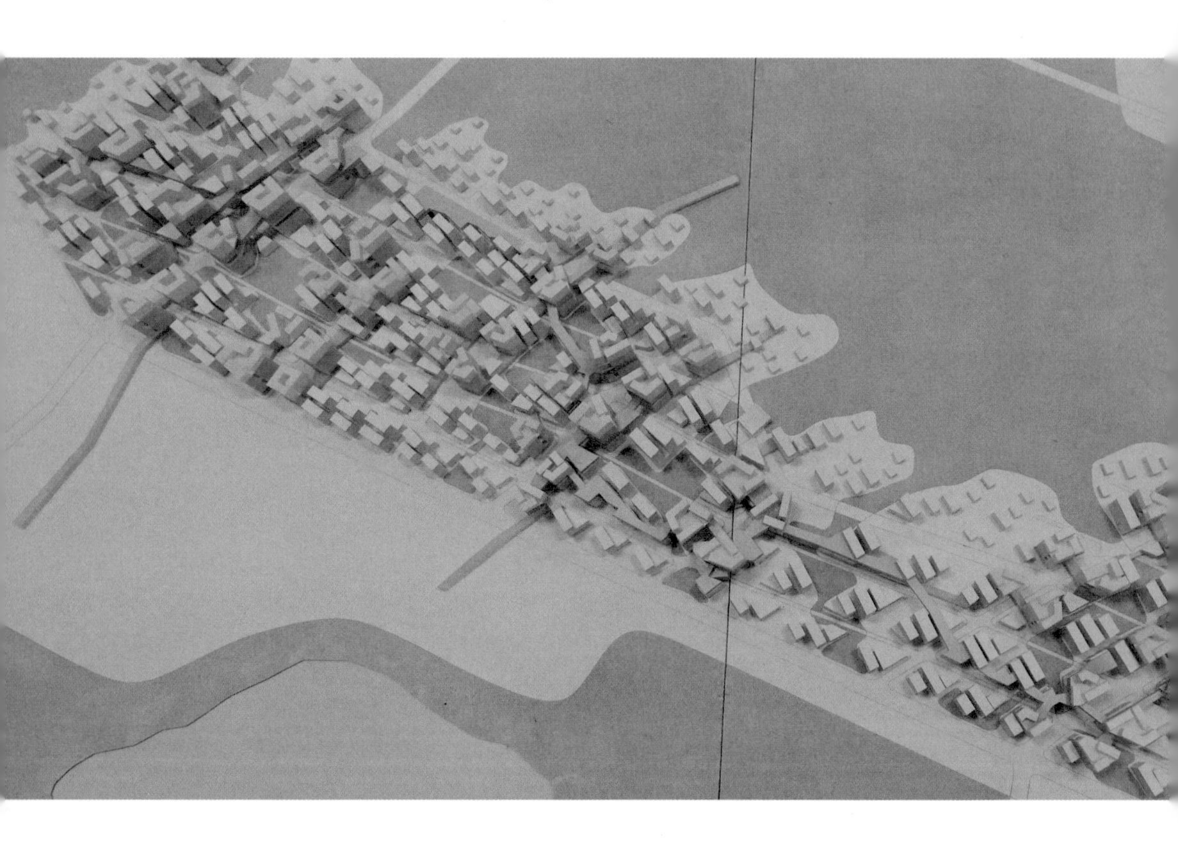

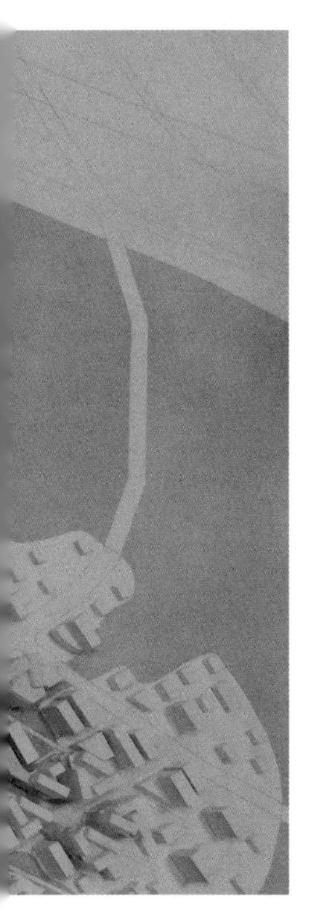

模型照片
Model Photo

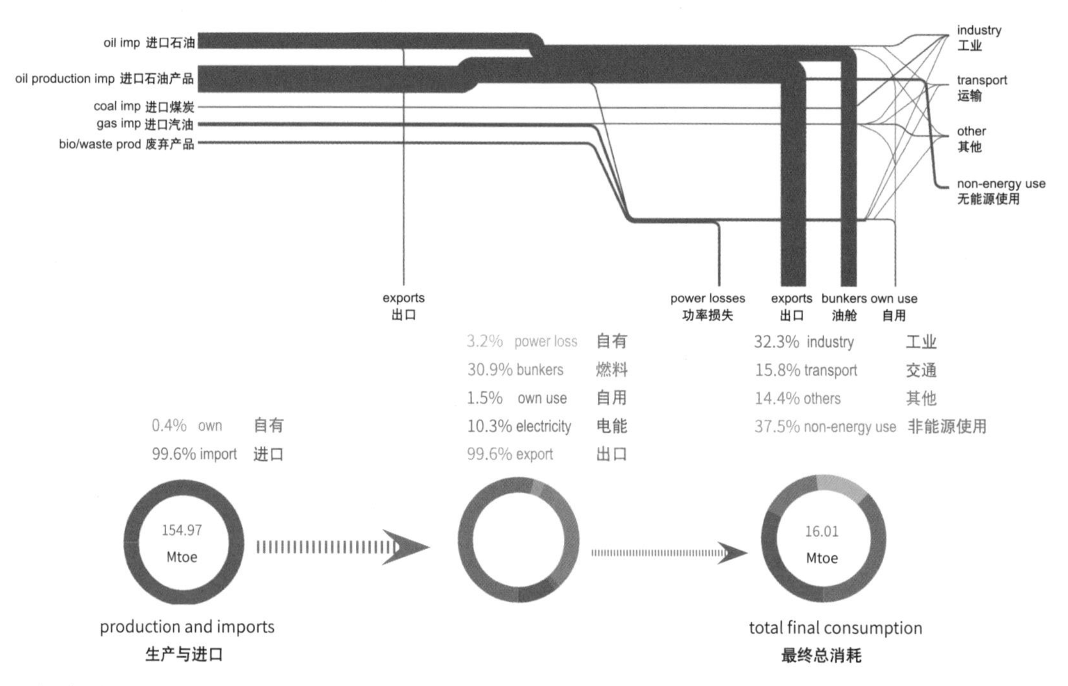


风险：资源流
RISK : Resource Flow

新加坡由于其色彩缤纷的热带气候而被称为“花园城市”，具有丰富的城市文化。到2020年，新加坡的人口将从500万增加到650万，这也是机遇和挑战并存的时期。

“风险”定义了新加坡将面临的四个未来挑战：资源、工业、社会和动力学。我们将使用“流”来解释其背后隐藏的逻辑。

Singapore, known as the “Garden City”,is a city with colorful urban culture thanks to its colorful tropical climate. Its population will increase from 5 million to 6.5 million by 2020, presenting both opportunities and challenges.

RISK defines four future challenges that Singapore will face: Resource, Industry, Society and Kinetic. We will use “flow” to interpret its hidden logic.

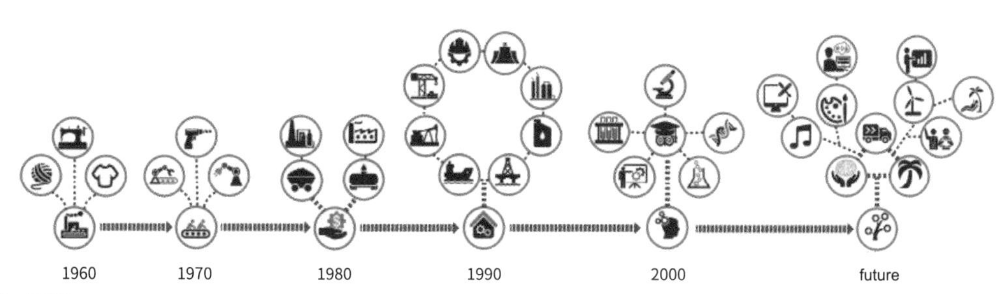


风险：工业流
RISK : Industry flow

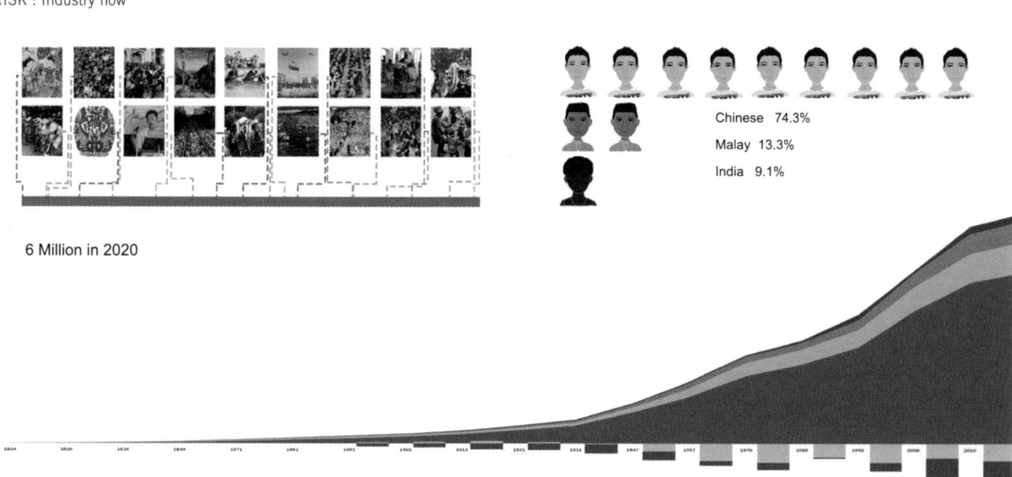


风险：社会流
RISK : Social Flow

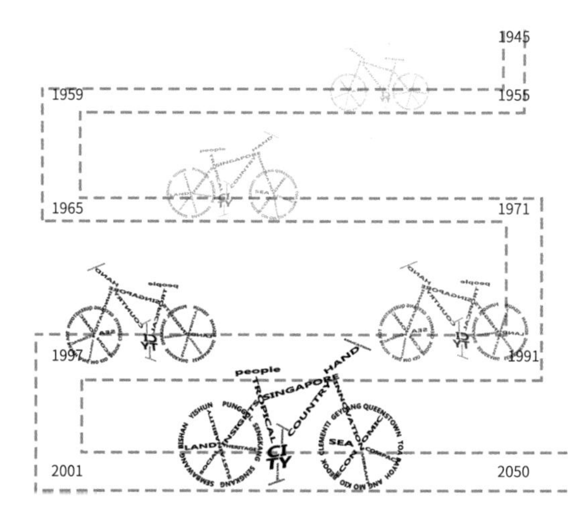


风险：动力学
RISK : Kinetic

新加坡就像一辆自行车。

新加坡的土地面积小，高效率的政府也具有灵活发展的优势，就像小巧轻便的自行车，在发展过程中不断调整。一旦停下来，它将面临衰退的风险，因此只有不断发展，就像一辆自行车只能向前推进，才能让新加坡走在世界的前沿。

Singapore is just like a bicycle.

As a compact and lightweight bicycle, small land area and high efficient government in Singapore have the advantage of flexible development and constantly adjustment. Once stop, it will face with the risk of decline. Only when it keeps moving, can it walk in the front of the world.

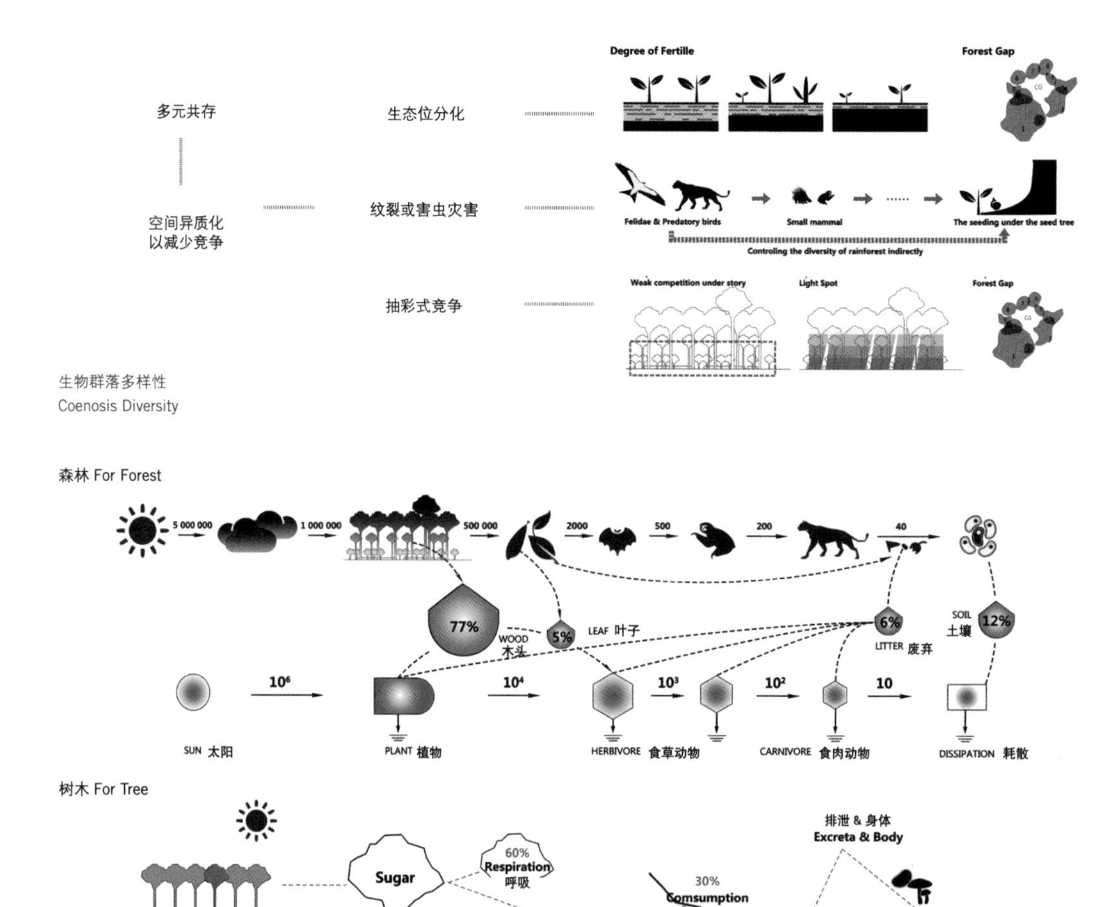


生物群落多样性
Coenosis Diversity

能量流
Energy Flow

新加坡位于赤道附近，其天气主要受热带雨林气候影响。高温多雨的天气、潮湿的热环境也对当地的传统建筑产生了巨大的影响。

我们向自然和传统学习，从雨林和传统的店屋中提取了自然原型，并以它们为基础进行了大量的热力学研究，从而确定了设计热力学城市的基本指南和原型。

Singapore is located near the equator and presents the features of the tropical rainforest climate. The high temperature, rainy weather, humid thermal environment also have a huge impact on traditional architecture.

We learned from nature and tradition, extracted the natural prototype from rainforest and traditional shophouse. A thorough thermodynamic study on them then lead us to the basic guide and prototypes to design the thermodynamic city.

建筑 Building

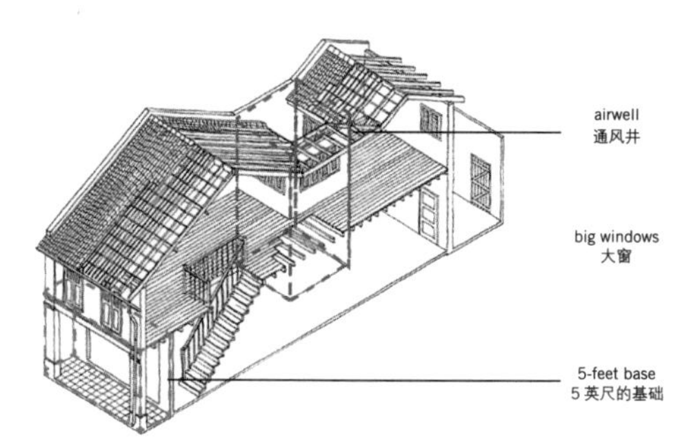


街区 Block

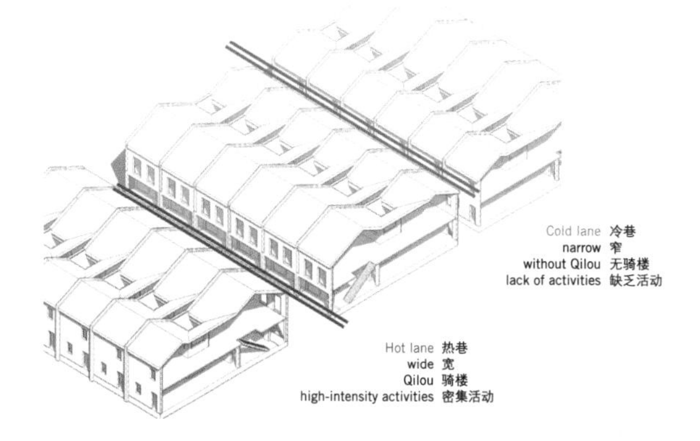


地区 Region

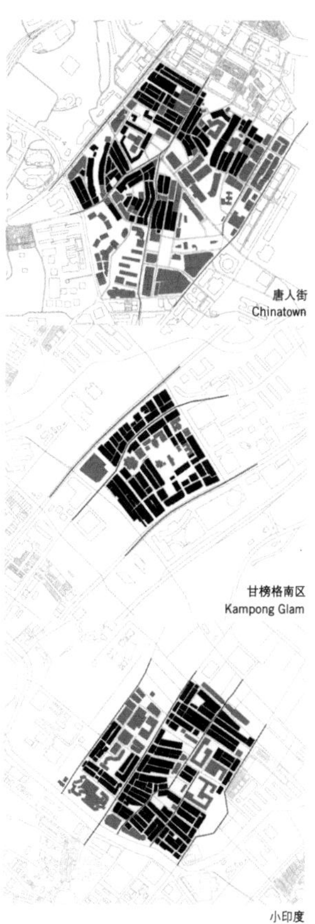

主街方向顺应盛行风向
Direction of Main Street Following Prevailing Wind

基于上述研究和理解，且因为能量流动涉及生活的各个方面，成果聚焦于能量流动并试图找到合理的流动模式。我们引入了基于当地气候、文化和挑战的热力学城市范式，即“雨林城市”。以此来解决潜在的挑战，并创造一个理想的城市。这里自然（空气、风、阳光、绿色等）更亲近人类，环境更舒适，每个个体都被紧凑的流动模式凝聚在一起。

雨林城市是一个新的城市模型。考虑到新加坡独特的气候和文化，我们提倡以“能量流”来组织社会联系，正如法国哲学家拉图尔所呼吁的那样，自然与社会之间应有更紧密的联系，以应对未来的城市挑战。

在热带雨林城市，我们以小而分散但凝聚的方式参与自然、工业、社区等的能源流动。我们将体验一个完全不同的城市，一个每个个体都共享和奉献的城市。

在热带雨林城市，“个体贡献”意味着每个个体对个人、家庭、社区、地区、城市和国家的责任感。每个人都参与不同层级的运营。同时，他们享受每个层级带来的好处。“每个个体贡献”都可以扩展到每个社区，每个社区之间相互平等的联系将创造一个同质、分散的城市。

Based on the above research and understanding, we focused on energy flow and tried to find a reasonable flow pattern, because energy flow is involved with all aspects of life. We introduced a thermodynamic urban paradigm based on the local climate, culture and challenges, a “Rainforest City”. We proposed “Rainforest City” to solve the potential challenges and to create an ideal city, where nature (air, wind, sunshine, green, etc.) is closer to human, environment is more comfortable, everyone is tightened by compact flow pattern.

Rainforest City is a new city model. Take the unique climate and culture in Singapore into account, we advocate the “energy flow”, as French philosopher Latour appeals, a closer connection between nature and society to meet the future urban challenge.

In the Rainforest City, we participate the energy flow of nature, industry, community, etc. in a small, scattered and gathering way. We would experience an entirely different city, a city where everyone shares and dedicates.

In the Rainforest City, “Everyone contributes” implies everyone's sense of responsibility for individual, family, community, region, city and country. Each person is involved in each level of operation. At the same time, they enjoy the benefits brought by each level. “Everyone contributes” can be extended to each community and the mutual equal connection among each community will create a homogeneous and disperse city.

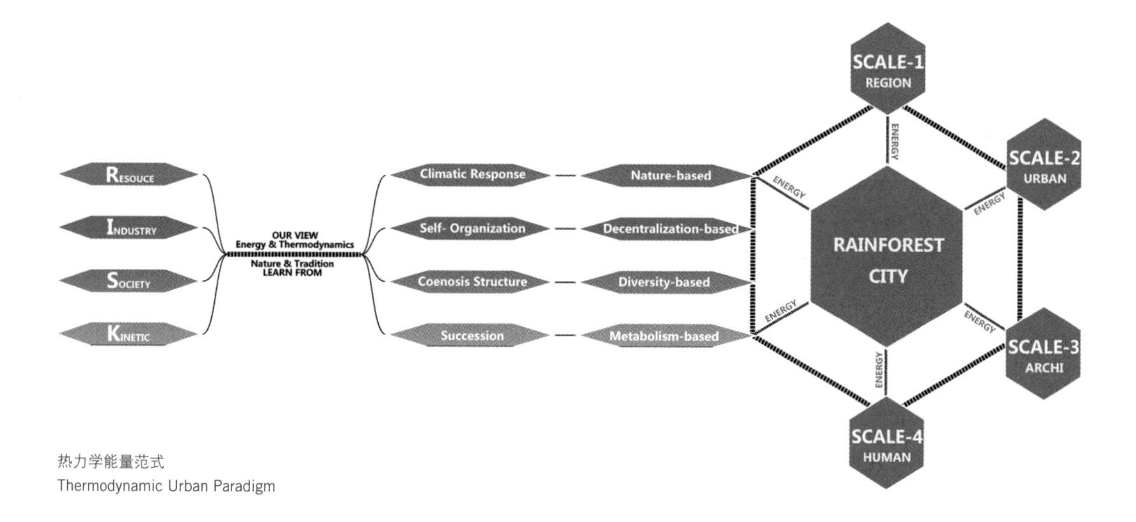


热力学能量范式
Thermodynamic Urban Paradigm

作为热力学原型，热带雨林城市拥有4个关键要素：

1.自然气候响应，风、空气、日照和雨水同时形成气流城市。

2.基于分权的自组织模式，引导能源流动城市实现高效的能源流动。

3.基于多样性的科学认知结构，构建人际流动的城市，人际关系密切。

4.以代谢为基础的城市演替，勾勒出灵活多变、可持续发展的城市流动。

As a prototype of thermodynamics, Rainforest City has 4 key elements:

1. The nature-based climate response, where wind, air, sunlight and rain shape the air flow city simultaneously.

2. The decentralization-based self-organization model guides the energy-flow city with high efficient energy flow.

3. The diversity-based coenosis structure constructs the population-flow city with close interpersonal relationship.

4. The metabolism-based urban succession outlines flexible, variable and sustainable developing city in flowing.

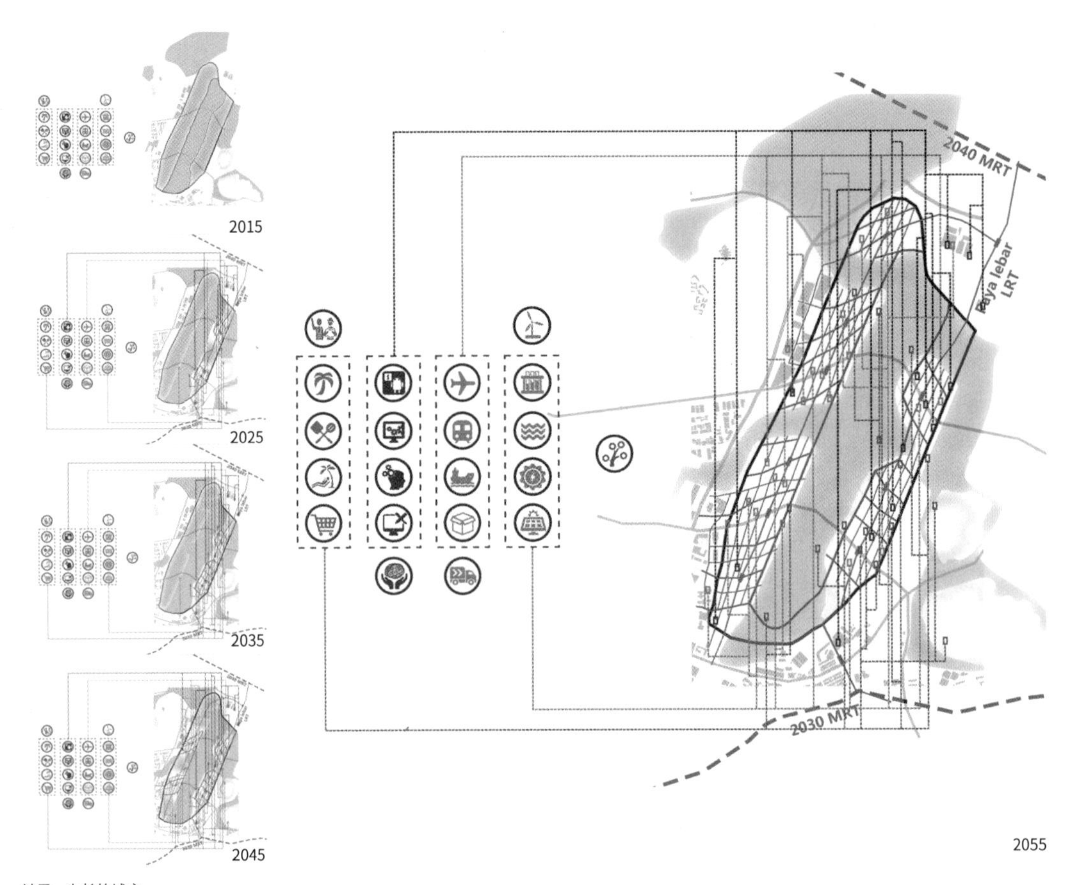


继承 - 生长的城市
SUCCESSION- Growing City

巴耶利巴空军基地是新加坡市中心最大的飞地。通过汽车、自行车或徒步来感知基地，我们发现，空军基地造成了城市社区的“孤立”。我们希望在地区的尺度内从三个不同方面——气候响应、产业循环和城市的继承来引导巴耶利巴发展到 2050年。

考虑到巴耶利巴的核心区位和规模，它的重建项目不仅是激活城市区域的机会，也是新加坡面对未来挑战的机遇。

形成以旅游业、创新&智能、物流和清洁能源行业为中心的高度混合型产业周期。

Paya Lebar Air Force Base is the largest enclave in the center of Singapore. After experiencing the base by car, by bicycle and on foot, we felt that the base has caused the isolation of urban communities. From the scale of region, we attempt to guide Paya Lebar to the year of 2050 in three aspects: climate responsive, industry circulation, and the succession of city.

Given its core location and scale, the redevelopment of Paya Lebar will present opportunities to activate urban areas and for the country to face future challenges.

We have formed a highly mixed industry cycle centered around tourism, innovation & intelligence, logistics, and clean energy industries.

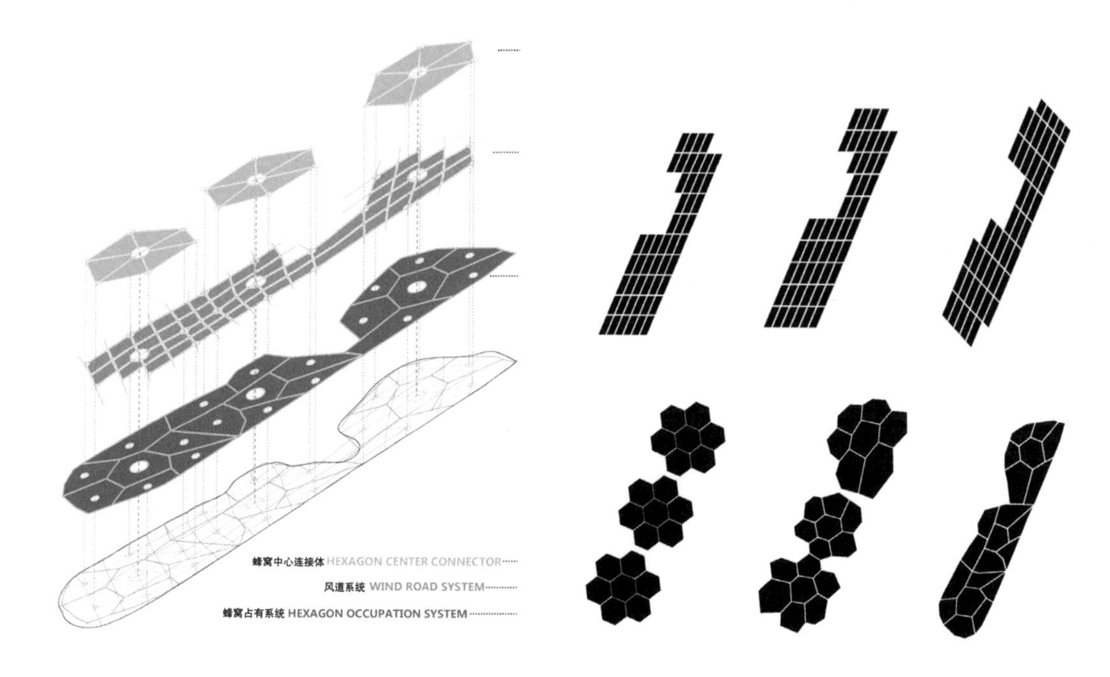


地域网络 Territory Net

道路网络 Road Net

一个可见的菱形网络和不可见的六边形空间应运而生
A visible diamond-shaped network and invisible hexagonal space came into being

根据巴耶利巴的整体规划，我们将能量流动策略应用于1平方公里的城市设计，以进一步发挥雨林城市的想象力。我们选择了森林和水库之间的狭长地形，并尽可能多地引入了自然因素。 这种设计基于气候来进行考虑。

更重要的是，我们更关注自组织无形的逻辑。为了使能量流动更加合理，我们引入六边形作为散射和聚集的社区场所。

Based on the overall planning of Paya Lebar, we apply the strategy of energy flow to the urban design of 1 $km^2$ for the further imagination of a rainforest city. A long and narrow terrain between the forest and the reservoir was selected and as many natural factors as possible were introduced to it. This design is based on the climate.

What's more, we pay more attention to self-organizing invisible logic. To make the energy flows more reasonably, we introduce hexagon as the community field which scatters and gathers.

## 热力学建筑原型 Thermodynamic Architectural Prototype

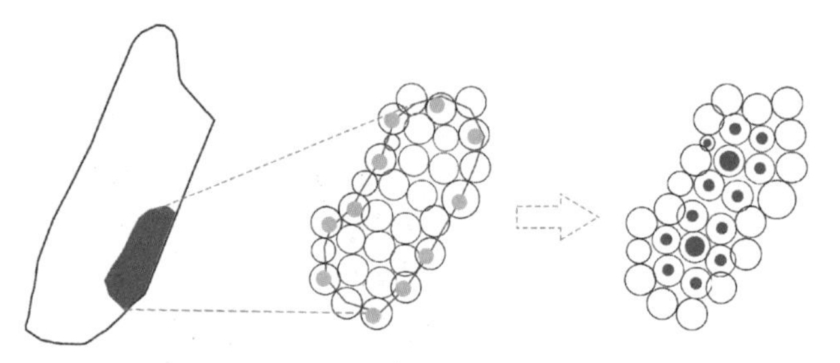

Selecting a part of our design to explain process. We conform several control points(such as contour, reservior, forest and so on) on the boundary line. Using physical simulation to generate the locations of hexagon.

六边形变形 Hexagon Deformation

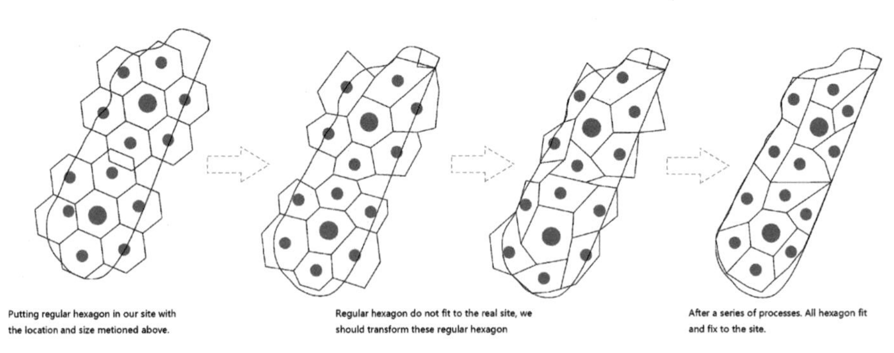

Putting regular hexagon in our site with the location and size metioned above.
Regular hexagon do not fit to the real site, we should transform these regular hexagon
After a series of processes. All hexagon fit and fix to the site.


六边形适应性变形 Adaptive Transformation of Hexagon

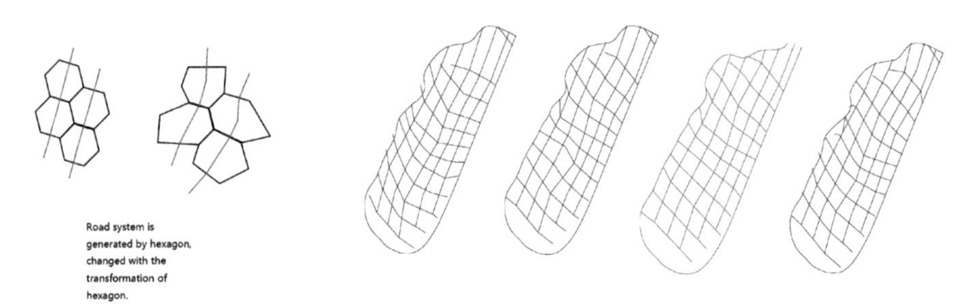

Road system is generated by hexagon, changed with the transformation of hexagon.


道路生成与优化 Road Generation and Optimization

六边形理论
HexagonTheory

通过数学分析比较六边形、三角形和矩形系统，结果表明六边形系统是提高整个系统效率的最佳选择。
在对案例进行分析后，考虑到轻轨步行距离、基本社会服务半径等因素，我们将六边形社区的面积确定为300m$^2$。

Comparision between hexagon, triangular and rectangular system through mathematical analysis shows that hexagon system is the best choice to improve the efficiency of the whole system.
After analyzing the cases, considering the LRT walking distance, radius of basic social services and so on, we identify the area of the hexagon community as 300m$^2$.

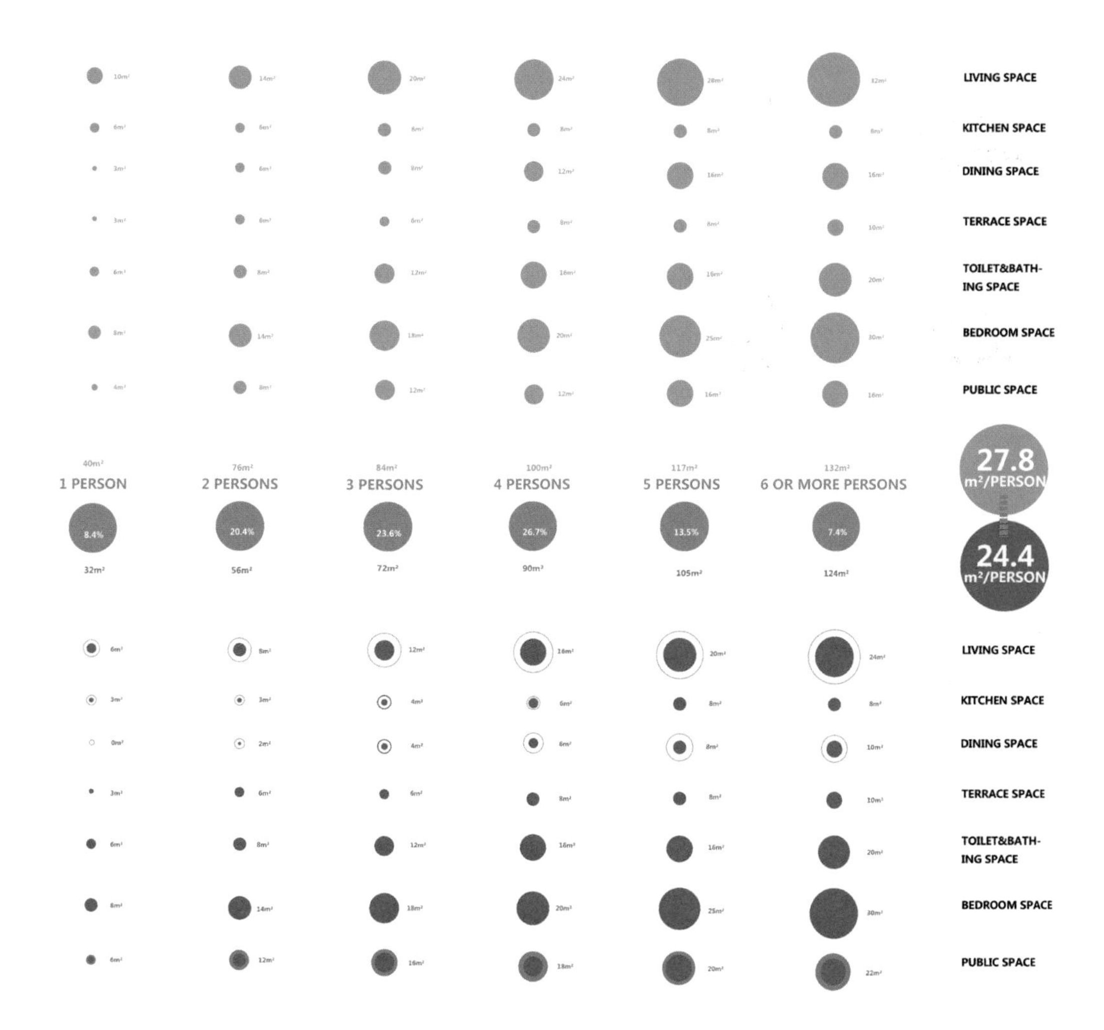


生物群落密度
Coenosis Density

分享是一种保持家庭亲密关系的方法，也是解决资源短缺问题的一种方法。

通过分享，每个人所需的平均空间减少到大约24.4到27.8平方米，一个家庭的总需求空间可以尽可能地压缩到87.76%。

因此，分享将是充分利用空间和其他事物的有效方式，有助于为每个人、每个家庭提供幸福的生活。

Sharing is a way to keep family close and a solution for the shortage of resouces.

By sharing ,the average space each person needs is cut down to about 24.4 to 27.8 square meters, and the total demand of space for a family could be compressed by 87.76%.

Therefore, sharing would be an effective way to make full use of space and all other things. This could help us to offer a happy life for every one and every family.

**热力学建筑原型 Thermodynamic Architectural Prototype**

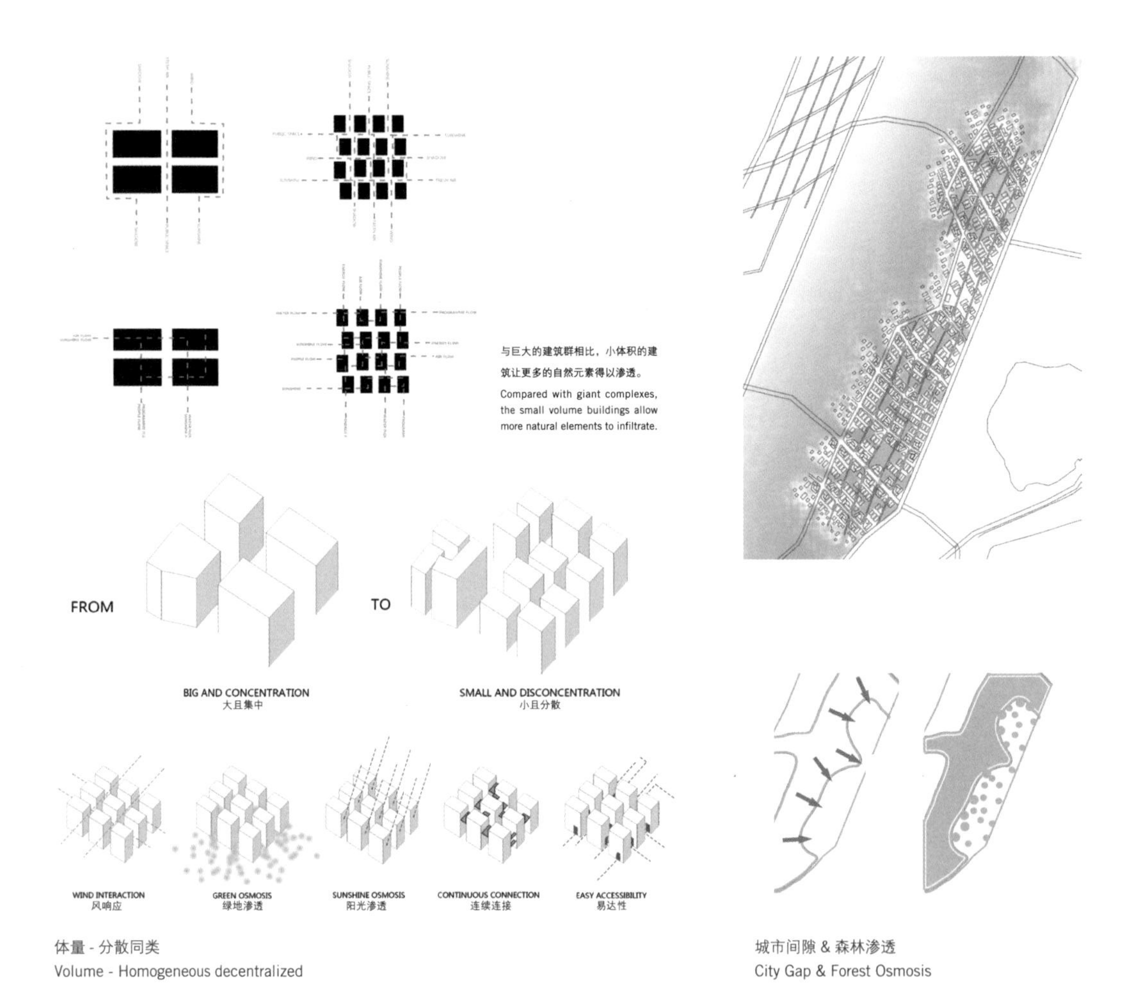


体量 - 分散同类
Volume - Homogeneous decentralized

城市间隙 & 森林渗透
City Gap & Forest Osmosis

城市冠层的韧性有助于空气更好地穿透。
因此，城市需要留下更多的空白，有节奏地安排公园和广场，以形成城市森林空白。

The resilience of urban canopy helps the air to penetrate better.
So the city needs to have more void, arrange Park and Plaza rhythmically to form urban forest gaps.

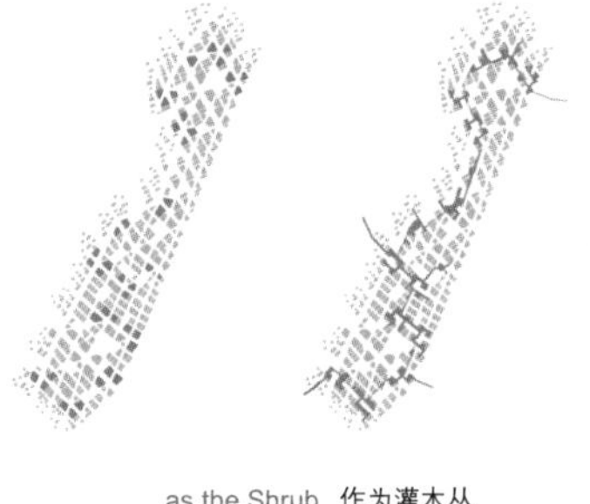

as the Shrub 作为灌木丛
green based minus space 基于绿植的负空间

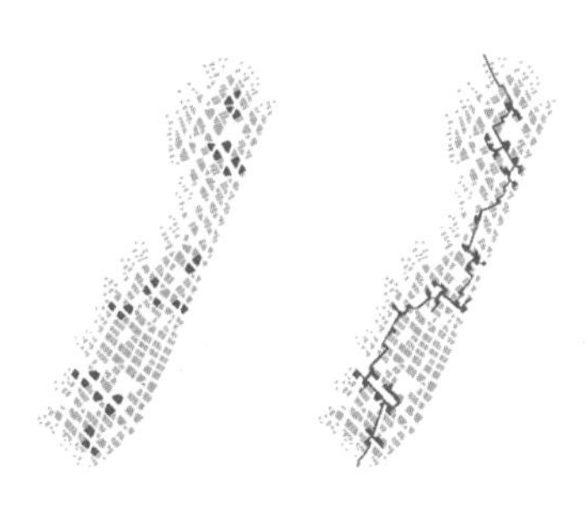

as the Canopy 作为树冠
street based public space 基于街道的公共空间

作为核心空间的街道
Street as Core Space

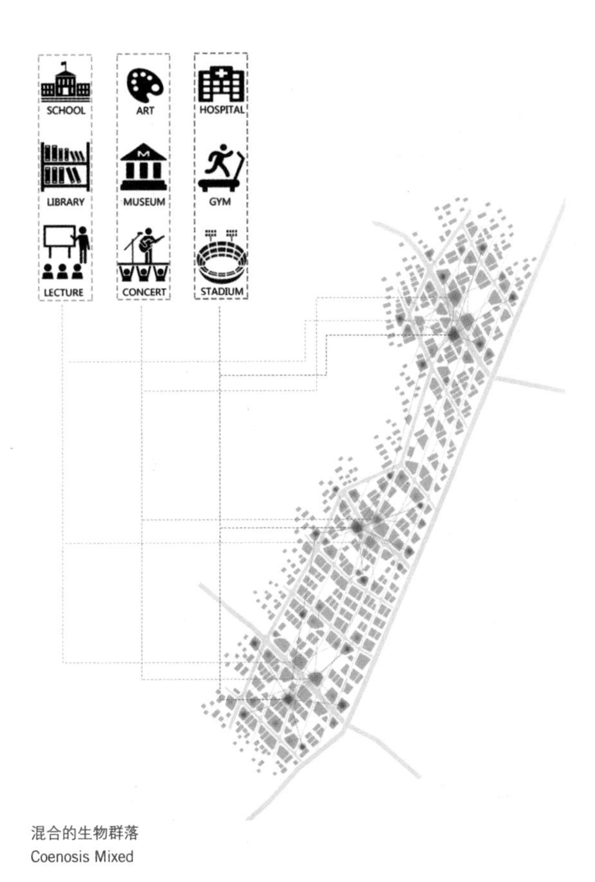


混合的生物群落
Coenosis Mixed

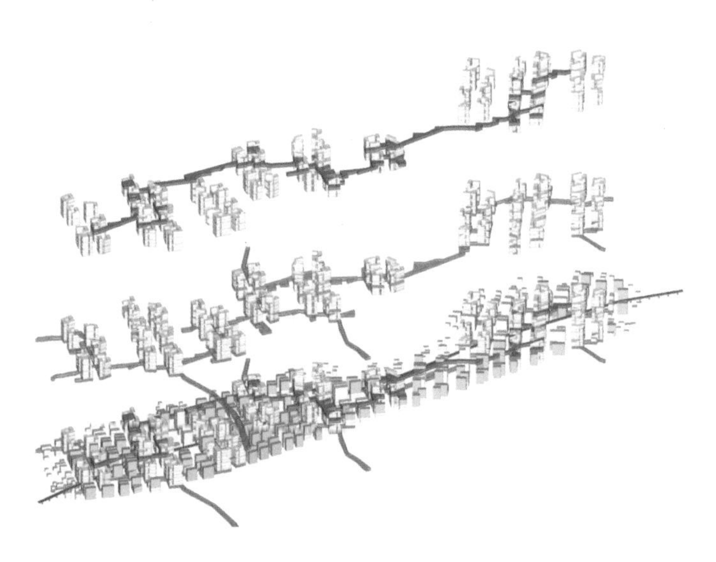

两个系统将小型建筑物紧密连接在一起，并帮助街道生活垂直向上发展，这也将社区活力从灌木层扩展到树冠层。

Two systems link the small volume buildings closely together, and help street life develop vertically above, which also extend the community vitality from the shrub layer to the canopy layer.

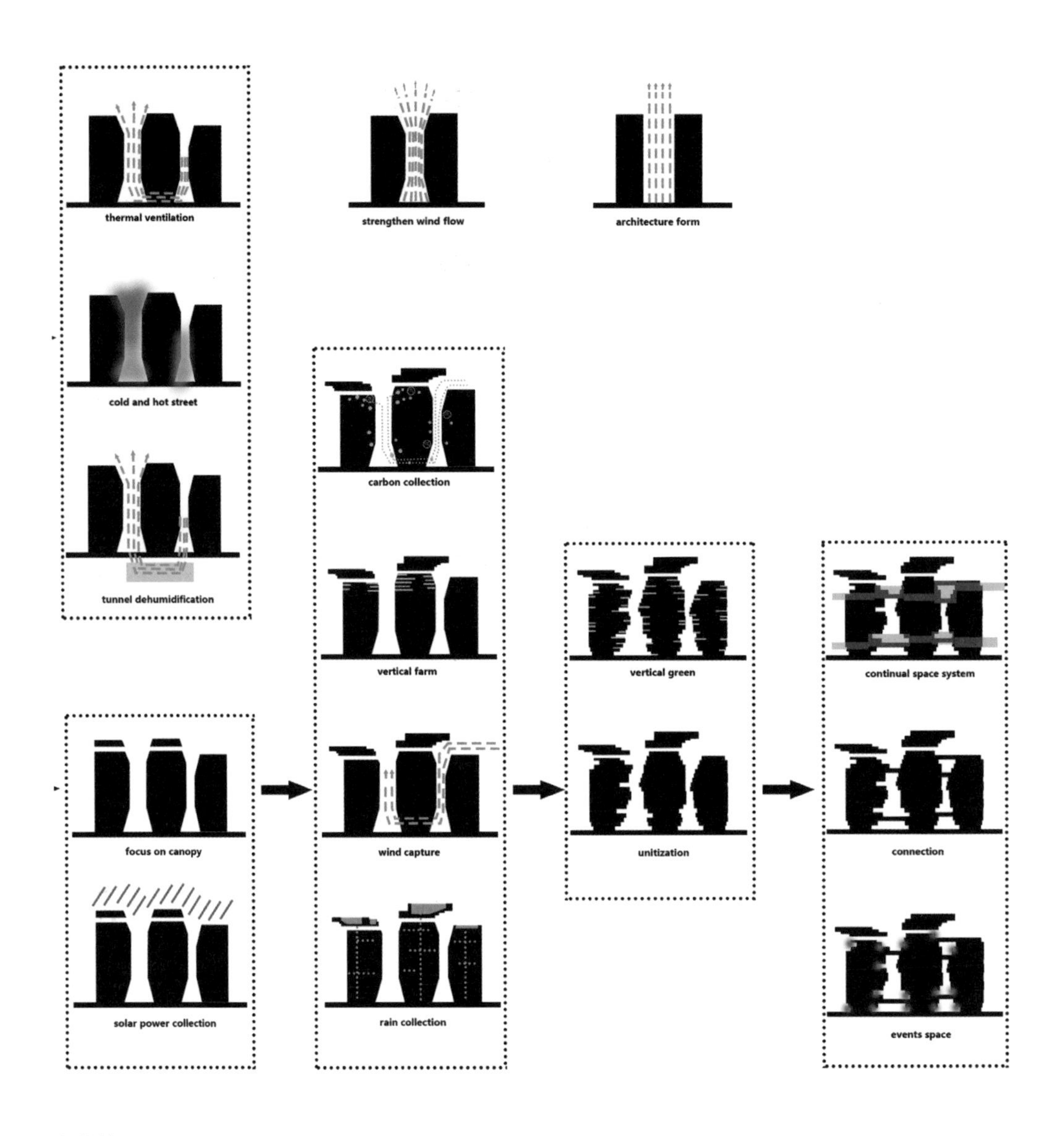


生成设计
Generating Design

我们提出了一种将自然元素与建筑、围绕建筑的生活相结合的建筑原型。作为建筑的重要产生器，自然元素和内部事件共同塑造了建筑形式。

We propose an architectural prototype which combines the natural elements with buildings and life with buildings. As the important generator of buildings, the natural elements and the internal events shape the building form together.

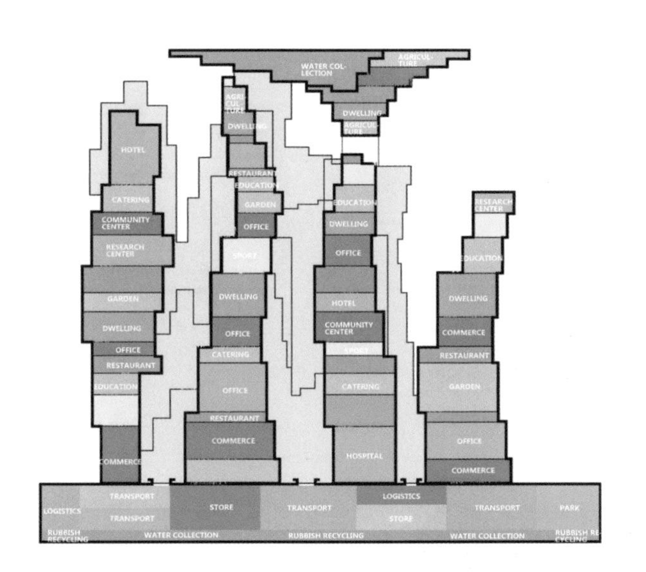


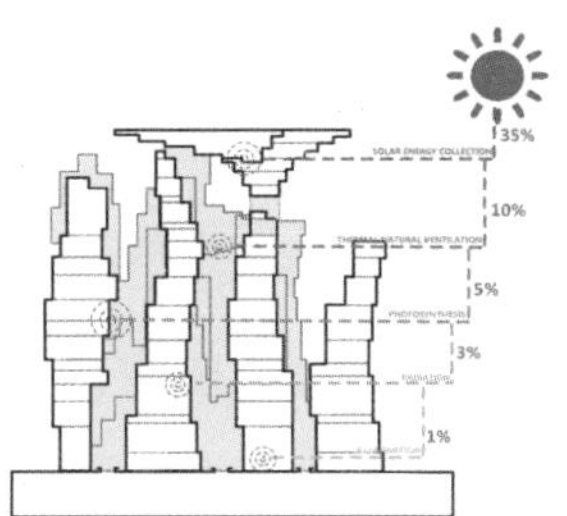


垂直结构
Vertical Structure

基于活动的循环
Events-based Circulation

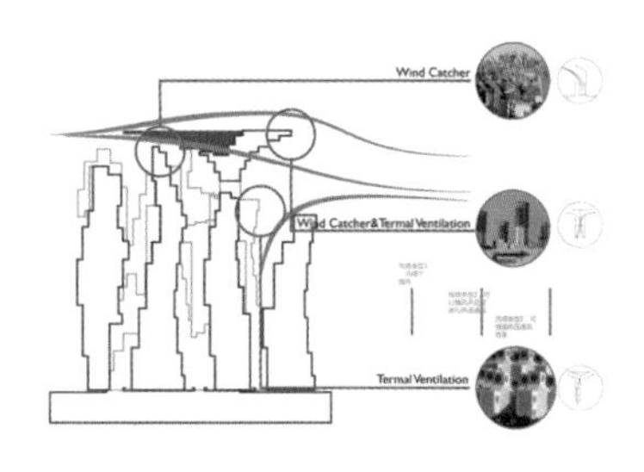


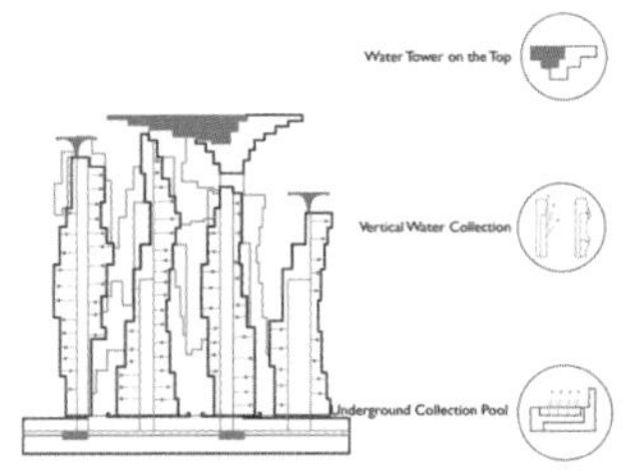


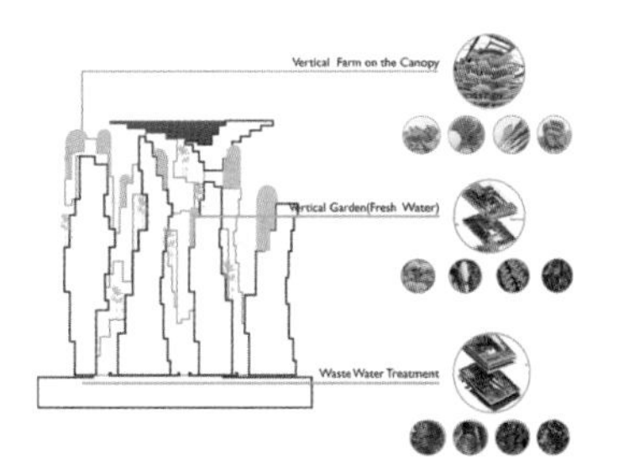


基于自然的发生器
Nature-based Generator

在我们的热带雨林城市里，不仅是建筑的内部空间，还有每一个建筑都为形成建筑间的城市空间做出了贡献。

In our Rainforest City, not only the internal space of architecture accounts, every architecture contributes to forming the urban spaces.

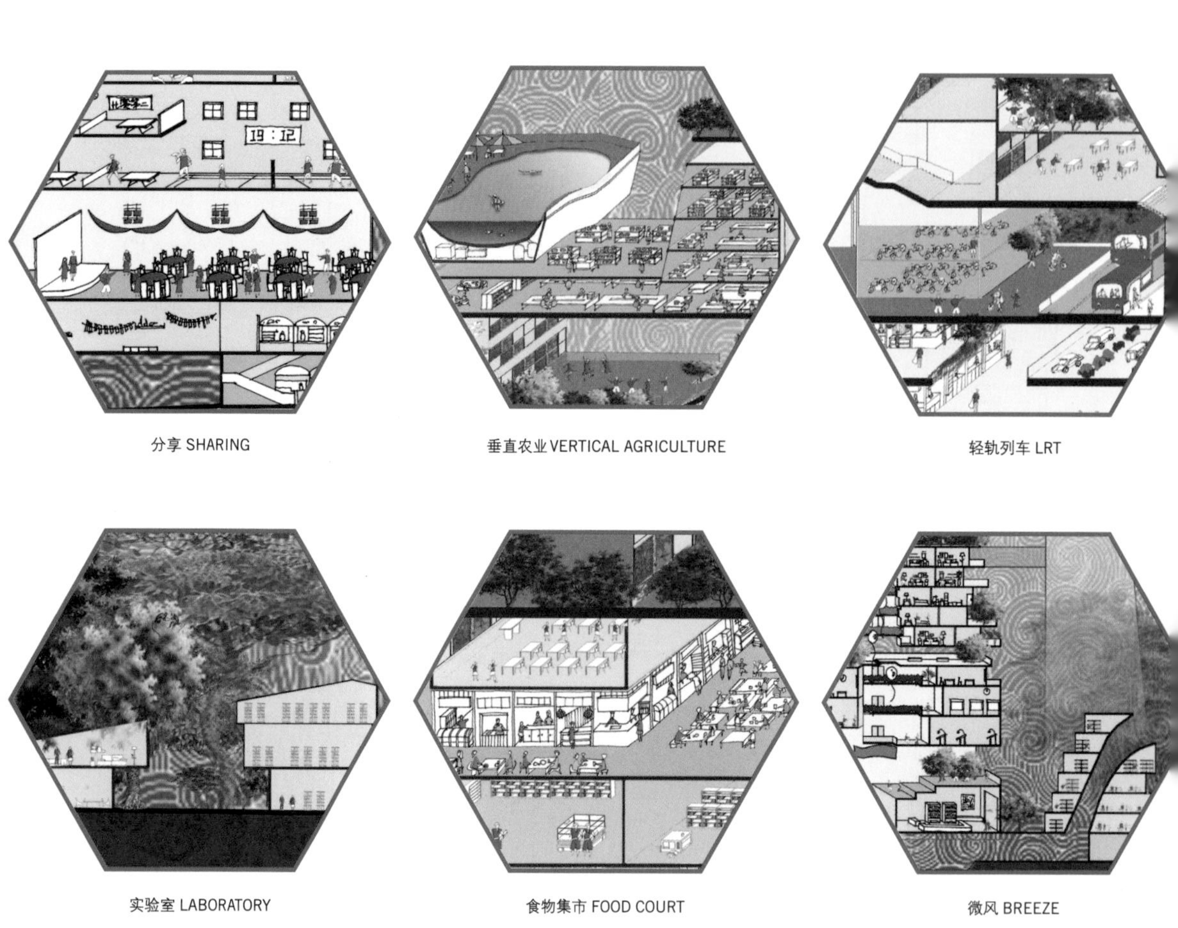

分享 SHARING

垂直农业 VERTICAL AGRICULTURE

轻轨列车 LRT

实验室 LABORATORY

食物集市 FOOD COURT

微风 BREEZE

人类 - 生活体验
HUMAN - Life Experience

当我们生活在热带雨林城市时，我们可以参与垂直生物群落的种植，与家人一起了解自然并做出贡献。 我们可以参加生物群落里的公共婚礼，与整个社区一起为新人祝福。

我们可以到生物群落形成的美食广场中，贡献和分享整个地区的美味佳肴。它是分享，但更是一种贡献。

When we live in the Rainforest City, we could participate in the vertical farming of cenosis, contribute and learn about the nature with our family. We could participate in the public wedding of cenosis, blessing the new couple with the whole community.

We could participate in the food court of coenosis, sharing the delicious food with the whole region. It is sharing, but also a contribution.

因此，在我们的热带雨林城市里：
每个个体意味着每个人、自然元素、城市的各个因素等。
每个个体都为城市运转和能量流动模式做出了贡献。同时，能量流模式以合理的方式组织每个个体，以便每个个体都能以自发有效的方式做出贡献。

Therefore, in our Rainforest City:
Everyone means every indivdial, every natural element, every factor of city and etc. Everyone contributes to the city operation and the energy flow pattern. Meanwhile, the energy flow pattern organizes in a reasonable way so that everyone could contribute in a spontaneous and effective way.

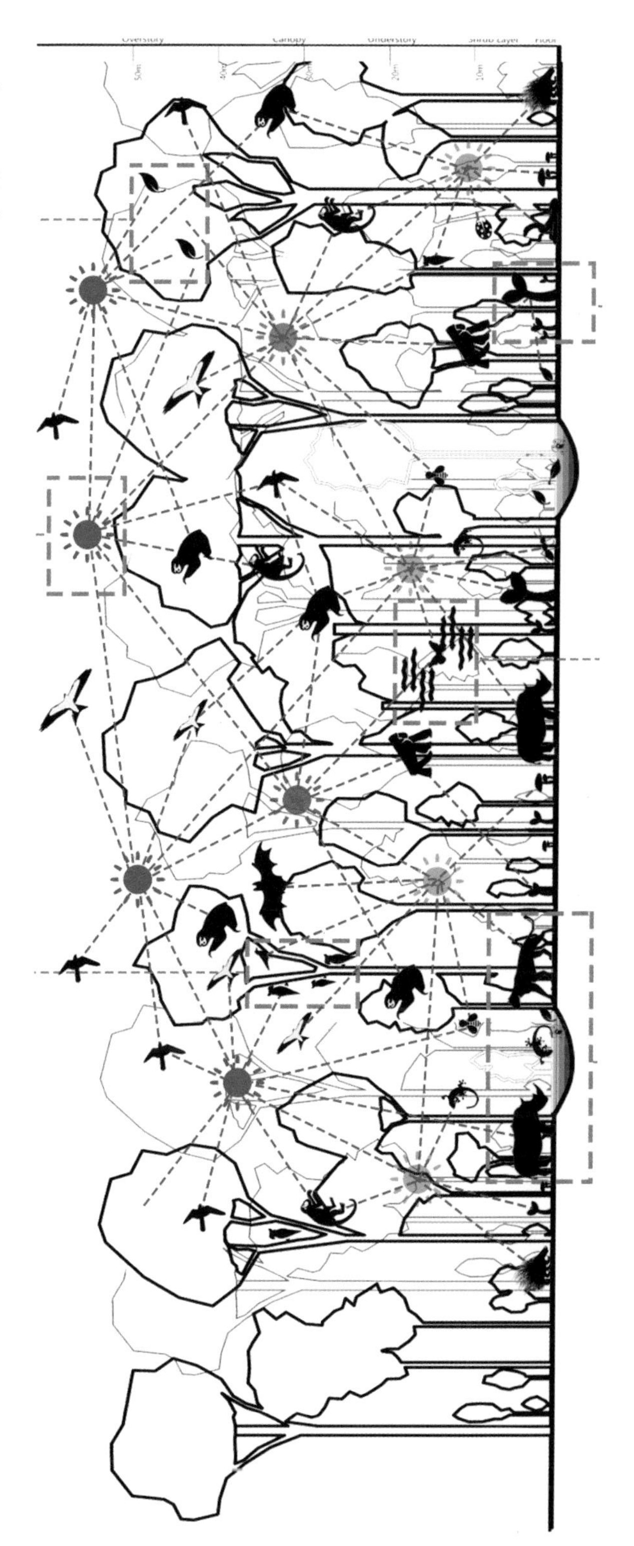

# 作者简介

**李振宇**

同济大学建筑与城市规划学院院长，教授、博士生导师。国务院学位委员会学科评议组建筑学学科成员，德国包豪斯基金会学术委员。

---

**李翔宁**

同济大学建筑与城市规划学院副院长，教授、博士生导师，长江学者青年学者，知名建筑理论家、评论家和策展人。

---

**李麟学**

同济大学建筑与城市学院教授、博士生导师，麟和建筑工作室 ATELIER L+ 主持建筑师，从 2018 年至今，同时担任同济大学艺术与传媒学院副院长（主持工作）。

---

**伊纳吉·阿巴罗斯**

哈佛大学设计研究生院建筑系前系主任，建筑系教授(2013—)；西班牙马德里建筑学院（ETSAM）博士（1991）与主持教授(2002—)；Ábalos+Sentkiewicz arquitectos 的创始人及主持建筑师。其作品以在建筑、环境与景观之间建立一种技术精确、形式想象与整体法则的初始综合而著名，可称之为“热力学美学”方法论。

---

**周渐佳**

上海冶是建筑工作室创始合伙人、主持建筑师，《时代建筑》杂志栏目主持人，同济大学建筑与城市规划学院客座讲师，香港大学建筑系上海中心讲师。

---

**谭峥**

同济大学助理教授，张永和教授“千人团队”教学科研执行人。加州大学洛杉矶分校建筑学博士，为该系首位华人博士学位获得者。长期研究与基础设施相关的城市形态与建筑策划，同时任《时代建筑》客座编辑与《新建筑》杂志审稿人。

---

**陈昊**

同济大学建筑与城市规划学院建筑学学士，哈佛大学设计研究生院城市设计硕士，HCCH STUDIO 主持建筑师。

---

**胡琛琛**

同济大学建筑与城市规划学院建筑学学士，哈佛大学设计研究生院城市设计硕士，HCCH STUDIO 主持建筑师。

---

**郑馨**

同济大学建筑与城市规划学院建筑学学士，哈佛大学设计研究生院建筑硕士在读。

---

**郑思尧**

同济大学建筑与城市规划学院建筑学学士，SCI-Arc 建筑硕士在读。

---

**吕欣欣**

同济大学建筑与城市规划学院建筑学学士。

---

**夏孔深**

同济大学建筑与城市规划学院工学学士、建筑学硕士，瑞典查尔姆斯理工大学建筑与城市设计硕士，现就职于同济大学建筑设计研究院（集团）有限公司原作设计工作室。

---

**范雅婷**

同济大学建筑与城市规划学院工学学士、建筑学硕士，夏威夷大学建筑学博士。

# About Authors

**LI Zhenyu**

LI Zhenyu is the dean, professor and Ph.D Supervisor at College of Architecture and Urban Planning Tongji University. He is a member of the Department of Architecture of the Disciplinary Review Group of the Academic Degrees Committee of the State Council and an academic member of the Bauhaus Foundation in Germany.

---

**LI Xiangning**

LI Xiangning is the deputy dean, professor and Ph.D Supervisor at College of Architecture and Urban Planning Tongji University.

---

**LI Linxue**

LI Linxue is a professor , Ph.D Supervisor at College of Architecture and Urban Planning Tongji University and the principal architect of ATELIER L+. He is the deputy dean presiding over work at College of Art and Media Tongji University from 2018.

---

**Iñaki Ábalos**

Iñaki Ábalos is the former Chair of the Department of Architecture and Professor in Residence since 2013 at Graduate School of Design Harvard University. He is a Ph.D. in Architecture (1991) and Chaired Professor of Architectural Design at the ETSAM since 2002. In association with Renata Sentkiewicz, he is a founding member of Ábalos+Sentkiewicz since 2006. The work they develop stands out for proposing an original synthesis of technical rigor, formal imagination and discipline integration between architecture, environment and landscape, an approach they have named "A Thermodynamic Beauty".

---

**ZHOU Jianjia**

ZHOU Jianjia is the co-founder and principal architect of YeArch Studio, coeditor of T+A Magazine, guest lecturer of College of Architecture and Urban Planning, Tongji University and Hong Kong University Shanghai Study Center.

---

**TAN Zheng**

TAN Zheng is Assistant Professor in architecture and urbanism at Tongji University. As an urban designer and architectural historian, Tan has taught at Los Angeles, Hong Kong and Shanghai. His research revolves around contemporary urban form with a focus on infrastructures and interiorized public spaces. Tan earned his Ph.D. in Architecture at the University of California, Los Angeles.

---

**CHEN Hao**

CHEN Hao is a Bachelor of Architecture from the College of Architecture and Urban Planning Tongji University and a Master of Urban Design from Harvard Graduate School of Design, the principal architect of HCCH STUDIO.

---

**HU Chenchen**

HU Chenchen is a Bachelor of Architecture from the College of Architecture and Urban Planning Tongji University and a Master of Urban Design from Harvard Graduate School of Design, the principal architect of HCCH STUDIO.

---

**ZHENG Xin**

ZHENG Xin is a Bachelor of Architecture from the College of Architecture and Urban Planning Tongji University and a Master of Architecture at Harvard Graduate School of Design.

---

**ZHENG Siyao**

ZHENG Siyao is a Bachelor of Architecture from the College of Architecture and Urban Planning Tongji University and a Master of Architecture at SCI-Arc.

---

**LYU Xinxin**

LYU Xinxin is a Bachelor of Architecture from the College of Architecture and Urban Planning Tongji University.

---

**XIA Kongshen**

XIA Kongshen is a Bachelor of Engineering and Master of Architecture from the College of Architecture and Urban Planning Tongji University. He is also a Master of Architecture and Urban Design from Chalmers University of Technology in Sweden. He is currently working at the Original Design Studio of the Tongji University Architectural Design and Research Institute.

---

**FAN Yating**

FAN Yating is a Bachelor of Engineering and Master of Architecture from the College of Architecture and Urban Planning Tongji University. She is a Ph.D. in architecture from the University of Hawaii.

# 主编简介

# Introduction of Chief Editor

**李麟学**

同济大学建筑与城市规划学院教授，博士生导师，麟和建筑工作室 ATELIER L+ 主持建筑师，能量与热力学建筑中心 CETA 主持人，《时代建筑》专栏主持人，哈佛大学 GSD 设计研究生院高级访问学者(2014)。2000 年曾入选法国总统交流项目“50 位建筑师在法国”，在巴黎建筑学院 PARIS-BELLEVILLE 学习交流。李麟学试图以明确的理论话语，确立建筑教学、研究、实践与国际交流的基础，将建筑学领域的“知识生产”与“建筑生产”贯通一体。主要的研究领域包括：热力学生态建筑、公共建筑集群、以及当代建筑实践前沿等。从 2018 年至今，同时担任同济大学艺术与传媒学院副院长(主持工作)。

李麟学主持建成或在建杭州市民中心(2013 世界高层建筑学会“世界最佳高层建筑”亚太区提名奖)、2010 中国上海世博会城市最佳实践区 B3 馆、四川国际网球中心、中国商贸博物馆、河南科技馆新馆等多项有影响力的建筑作品。曾获得中国建筑学会“青年建筑师奖”(2006)，上海青年建筑师“新秀奖”(2005)，“同济八骏”中生代建筑师(2017)，上海市建筑学会首届“上海市杰出中青年建筑师”(2018)等荣誉，获得国内外各类专业设计奖项二十余项。 曾参加“40 位小于四十岁的华人建筑师设计作品展”、“从研究到实践”米兰建筑三年展(2012)、威尼斯建筑学院国际工作室特邀教授(2013)、深港城市建筑双城双年展(2013)、上海城市空间艺术季城市更新展(2015)、“走向批判的实用主义：当代中国建筑”(哈佛 GSD,2016)等展览、论坛与学术活动。主持国家自然科学基金资助项目“基于生态化模拟的城市高层建筑综合体被动式设计体系研究”“能量与热力学建筑前沿理论建构”、上海市科委“环境性能与热力学导向的崇明生态岛低能耗建筑整合设计研究”等重要课题，在国内外核心专业刊物发表论文六十余篇，主编《热力学建筑视野下的空气提案：设计应对雾霾》，客座主编时代建筑《形式追随能量：热力学作为建筑设计的引擎》。基于其“自然系统建构”的建筑哲学与创造性实践，成为中国当代建筑的出色诠释者之一，也是国际学术领域热力学建筑与生态公共建筑集群的积极推动者。

**LI Linxue**

Prof. Dr. LI Linxue Li is a Ph.D. Supervisor at College of Architecture and Urban Planning of Tongji University, principal architect of ATELIER L+, director of CETA (Center for Energy & Thermodynamic Architecture), chairman of special column in *T+A*, and visiting scholar at Graduate School of Design Harvard University in 2014. In 2000, he was selected by the Presidential Program “50 ARCHITECTES EN FRANCE” and studied in Ecole d'Architecture de Paris-Belleville. LI Linxue tries to establish his architectural teaching, research, practice and international exchange based on the definite theoretical discourse and to integrate the production of knowledge and production of buildings in the field of architecture. His main fields of researches include thermodynamic ecological architecture, public architectural conglomerate, and frontier for contemporary architectural practice ect.. He is the deputy dean presiding over work at College of Art and Media Tongji University from 2018.

LI Linxue is responsible architect for many influential built or under construction projects such as Civic Center of Hangzhou which is 2013 Best Tall building Norminee of Asia & Australasia Region from CTBUH, Hall B-3 in Urban Best Practice Area of EXPO 2010 Shanghai China, Sichuan International Tennis Center, China Commerce and Trade Museum and New Henan Technology & Science Museum etc. He is honored with Young Architect Award (2006) by Architectural Society of China, Shanghai Young Architect Rookie Award (2005), Architects of Middle-aged Generation from Tongji (2017), First Shanghai Outstanding Young Architect (2018) by Architectural Society of Shanghai, and more than thirty international and domestic professional design prizes. He attended 40 Under 40 Exhibition, From Research to Practice (Milan Triennal, 2012), IUAV Workshop as a guest professor (2013), Bi-city Biennial of Urbanism/ Architecture-Shenzhen (2013), Shanghai Urban Space Art Season-Urban Regeneration (2015), and Towards A Critical Pragmatism: Contemporary Architecture in China (GSD Harvard,2016)etc. Meanwhile he has charged important programs like the research of Passive Design System for Urban High-rise Building Complex Based on the Ecological Simulation financed by National Natural Science Foundation of China, Construction of the Advanced Theory for Energy & Thermodynamic Architecture, and Study on Integral Design of Low-Energy Buildings on Chongming Eco-island Oriented by Environmental Performance and Thermodynamics financed by Shanghai science and technology commission ect. He's authored more than sixty papers in both international and national core journals. He has edited *Air through the Lens of Thermodynamic Architecture: DESIGN AGAINST SMOG, and coedited Form Follows Energy: Thermodynamics as The Engine of Architectural Design* for *T+A*. Linxue Li has the honor to become one of the excellent interpreters for contemporary Chinese architecture through his NATURE BASED SYSTEM architectural philosophy and creative practice. He is also an active promoter for thermodynamic architecture and ecological public architectural conglomerates in the international academic field.